Paso Robles
An American Terroir

By

Thomas J. Rice, Ph.D.

AND

Tracy G. Cervellone, C.W.E.

Published by the Authors
Paso robles, CA

Thomas J Rice, Ph.D.

Copyright © 2007 by Thomas J. Rice., Ph.D. and Tracy G. Cervellone, C.W.E.

Library of Congress Cataloging-in-Publication Data

Paso Robles: An American Terroir

by Thomas J. Rice., Ph.D. and Tracy G. Cervellone, C.W.E.

Includes bibliographical references, original photographs and index.

ISBN-13: 978-0-9799406-1-3

ISBN-10: 0-9799406-1-3

1. Paso Robles AVA. 2. Soil science and geology. 3. Wineries. 4. Viticulture

About the Authors

Dr. Thomas J. Rice is a Professor of Soil Science in the Earth & Soil Sciences Department, California Polytechnic State University (Cal Poly), San Luis Obispo. Dr. Rice is a native of Marshfield, Wisconsin. He earned a B.S. degree in Natural Resources from the University of Wisconsin, Madison in 1974, a M.S. degree in Soil Science from Montana State University, Bozeman in 1976, and a Ph.D. degree in Soil Science, Geology minor, from North Carolina State University, Raleigh in 1981. He has been a California Polytechnic State University faculty member since 1981, where he is responsible for teaching university courses in soil science, land use planning, soil geomorphology, soil resource inventory, and advanced land management. Dr. Rice has been a Certified Professional Soil Scientist (C.P.S.S.) since 1982. He has directed and performed soil resource inventories in California, Montana, North Carolina, Utah and Wisconsin. He has directed and produced over 15 soil survey reports in the Paso Robles American Viticultural Area since 1992. He has supervised over fifty master's theses and senior projects dealing with soil chemistry and viticulture, soil resource inventories of vineyards, soil-serpentinite mineralogy relationships, environmental mercury contamination, land use planning, and nonpoint source pollution-water quality studies. Dr. Rice has published numerous journal articles, research reports, and popular press articles. He has been the project director for funded studies involving soil and wine grape quality relationships, soil taxonomy updates in California, Nevada, and Utah; a comprehensive soils database for California; nonpoint source pollution in western rangelands; mercury pollution in a California watershed; soil mapping of a national wildlife refuge; and California soil map unit interpretation record updates. Dr. Rice serves as a private soil science consultant and has testified as an expert witness in legal depositions, California civil courts, and legal arbitration proceedings. Email: *trice@calpoly.edu* or *pasoterroir@yahoo.com*

Tracy G. Cervellone, C.W.E. is a native of Colorado. She has been employed as a salesperson in the wholesale wine business throughout California for over 20 years. Her love affair with the Paso Robles AVA and its wines dates back to the 1980's. A Certified Wine Educator (C.W.E.) since 2005, she also holds the "Advanced Certificate with Distinction" from the Wine and Spirits Education Trust (WSET). Currently, she is studying for the rigorous Master of Wine (M.W.) certification. She resides in southern California with her husband, Michael, and an assortment of dogs, cats, and fish. Tracy is an avid free diver, gardener, and self-confessed history, literature, philosophy and muscle car buff. She is the author and publisher of another year-2007 literary work, "Mommy Nation: the Decline of Dignity, Common Sense and Self Reliance in our Everyday Lives." Tracy also writes a regular national column for "Dining Out" magazine. Email: *cervellone@cox.net*

CONTENTS

Writing a book is an interesting, fascinating, and sometimes challenging exercise. We have met many enthusiastic and cooperative folks during book production and have learned that even the wildest dreams can come true.

- Paso Robles: An American Terroir
  This book is the interwoven story of the human history of the Paso Robles AVA along with its natural history. It is a book written to document the cultural history of many winery owners in the Paso Robles AVA, together with factual writings about the natural environment, including the geology, soils and climatic conditions of this area.

- Who are our readers?
  We believe that our initial audience will likely be those winery and vineyard owners who have made Paso Robles their home. Another group of readers will be those persons who visit the Paso Robles area for wine touring, are interested in learning more about the environment they experience, and enjoy the winery people that they meet. Internationally, wine connoisseurs who enjoy Paso wines will be able to learn more about this area's climate, land and people.

- Who is our publisher?
  After much thought and discussion, we decided to self-publish this first edition. Computers, printers and publishing software available today enable nearly anyone interested in learning the software programs and investing in the computer hardware the same rewarding opportunity.

- Book organization and companion CD
  We organized this book to give the reader an opportunity to gain an understanding of the Paso Robles AVA natural environment prior to reading the several winery stories. The winery stories are organized into chapters, which were named according to the eleven districts recently proposed by the Paso Robles AVA Committee. Appendices provide a soil survey report case study of a representative vineyard, information about soil properties related to viticulture, a complete directory of the area's wineries, and useful maps of the Paso Robles AVA. The index will assist the reader in efficiently locating information within the book.

  A companion CD, separately available for purchase via our web page, contains a PDF file of the entire book, along with web-linkable URL's and email addresses for every winery in the Paso Robles AVA. Additional JPEG images of geologic formations, soils, landscapes, vineyards and wineries are also included on the CD, organized by subject matter.

- Selection of winery stories and future edition
  This first selection of the winery stories was difficult, since many of the Paso Robles AVA wineries share great qualities. There simply was not room for the nearly 200 stories; a line had to be drawn. After four years in earnest production, time was running out. We needed to deliver this book to our readers. We wanted this first edition to be about real people, the folks that brought Paso Robles to the dance, and who both own wineries and have wine tasting rooms. There will be a revised and updated version in the future. Some stories will remain, most will evolve and new ones will be added. In Appendix C, we are as complete and accurate as possible to include all the Paso Robles AVA's wineries and their contact information.

  We welcome your comments and suggestions. Please contact us if there are errors of omission or factual misstatements. We can be reached via our web site or by email. We hope that you enjoy this book and will build it into your own account of this world-class wine region, the Paso Robles AVA.

  URL: www.pasoterroir.com
  Email: pasoterroir@yahoo.com

ACKNOWLEDGMENTS

Dr. Tom Rice is a very patient man, thank heaven. He was always supportive and helpful during the torturous process of condensing decades of learning and experience into a few hundred pages. I'm in awe of the depth and breadth of his knowledge, and equally and eternally grateful for his calmness and positivity, without which I would have floundered ceaselessly in a lake of winery stories. He has helped me to become a better writer, a better student and even a better person. I owe him a debt that I can never repay.

How do you thank a thousand people at once? There aren't words to express my gratitude for the open hearts, hands and wineries of all our friends in Paso Robles. Over the course of hundreds of contacts, interviews, emails, voicemails and visits over the last four years, you continue to inspire us with classic American diversity, fantastic stories, and "can do" attitude. In particular, we would like to acknowledge the assistance of the Paso Robles Wine Country Alliance, whose involvement, both directly and indirectly, was key to the success of this book.

Our thanks to Anthony Bowers for his photography contributions and continued support. And, to our long suffering spouses, Michael Cervellone and Nancy L. Jennejohn, who have put up with this challenging endeavor with grace and cheer. My thanks to the Arciero family, Kerry and Chris Vix, and the whole team at EOS Estate, for allowing me to distribute their wonderful wines to clientele for over four years and, in the process, fall in love with Paso Robles. To Luis Cota, one of my first real mentors, thank you for supporting all of my efforts to learn and develop in the wild wine business we have chosen. To my mentor Peter Koff MW, thank you for being patient while my Master of Wine studies and essays suffered for the last four years. Lastly, my thanks to Vern Underwood, Dick Maher, Andy Fromm and Mark Sneed, and the teams at Wilson Daniels and Young's Market Company, who have directed my career in such a positive way for a decade now: simply put, the boat doesn't float without you. To the legion of our professional and personal associates who supported this unique project with such enthusiasm, cheers and God bless to every single one of you.

This book is only a snapshot of a rapidly growing appellation. It will be an ongoing project, as Paso Robles' wineries and vineyards continue to evolve. New wineries are always starting up, some of which didn't make it into this edition, but will be included in the next update. The wine industry is as fluid as wine itself: living, breathing, changing with every sunrise. We look forward to exploring the future of Paso Robles as one of the world's best fine wine *terroirs. Cent' Anni*...(for a hundred years).

Tracy G. Cervellone, CWE
Lake Forest, CA

Many folks, too numerous to list, have been involved in the production of this book. However, many persons need specific mention. My first thank-you goes to my coauthor, Tracy G. Cervellone, who has persevered through many years of travels and sacrifices to complete her writings. My wife, Nancy L. Jennejohn, deserves my deepest gratitude and love for her encouragement and patience. My life would not be possible without the enduring love of my parents, Dr. Tom and Carol Rice.

There has been considerable involvement of many persons in the Paso Robles AVA who deserve special consideration. Steve Vierra, soil scientist and former student of mine, has shared many of his regional soil mapping experiences with me. He has been an invaluable source of local vineyard soils information for my Cal Poly students and me. The Carmody McKnight family, including Gary Carmody Conway, Marian McKnight Conway, Kathleen Conway, and Greg Cropper, have graciously allowed my students and I access to their vineyards for several studies of soils and vineyard relationships. They have become our dear friends over these many years of scientific discovery. Justin and Deborah Baldwin have been most cooperative during my early years of vineyard soil mapping. Our first Cal Poly vineyard soil maps were completed at JUSTIN vineyards in the early 1990's. Juan Nevarez shared many memorable field days in the vineyards with me during his early years at Carmody McKnight. It has been very exciting to watch Juan and his family succeed as vineyard managers and owners throughout this region. Mitch and Leslie Wyss have generously opened the gates of Halter Ranch for my Cal Poly students to conduct soil surveys, classroom soils projects, and a national collegiate soils contest. Many new academic and personal discoveries were made possible due to their assistance. Steve Glossner, a terrific enologist, has shared his winemaking knowledge with many Cal Poly students during our soil mapping projects at Justin Vineyard, Adelaida Cellars lands (HMR and Viking vineyards) and Halter Ranch. We are fortunate that he is now regularly teaching the Cal Poly introductory enology courses. Dick Hoenisch, while manager at Tablas Creek Vineyard, stimulated several Cal Poly senior projects regarding soil and rock chemistry of earth materials from the Monterey Formation. Thanks to Dick and to Bob Haas, Tablas Creek Vineyard co-owner, for their kind assistance. Elizabeth Van Steenwyk, proprietor, and Paul Sowerby, sales manager, of Adelaida Cellars facilitated many years of Cal Poly soil mapping projects on the HMR and Viking vineyards and surrounding lands. Many Cal Poly students have gained employment in the vineyard and wine business due to their encouragement. Mark Goldberg and Maggie D'Ambrosia of Windward Vineyard have been most helpful in opening their Pinot Noir vineyard for the production of a Cal Poly senior project about vineyard soils. They are precious friends and I never tire listening to Mark extolling the virtues of Pinot Noir. Mary Baker and Dan Panico of Dover Canyon Winery have been steadfast supporters of our efforts to better understand the soils and wines of the Paso Robles area. Mary is a prolific writer, who maintains the Paso Robles AVA Appellation America web site and an exciting wine blog. Thanks to Mary for reviewing this book and for her helpful suggestions. I treasure the memories of our many conversations about wines, soils and geology. Matt Trevisan, Cal Poly biochemistry graduate and proprietor of the family-owned winery Linne Calodo, has been a long-time proponent of the importance of vineyard *terroir* to ultimate wine quality. The name of his winery was chosen from two of the important vineyard soil series with the Paso Robles AVA. Gary Eberle of the pioneering Eberle Winery has long been interested in the earth science aspects of wine grape production. Numerous conversations with Gary have stimulated me to better understand the Paso Robles AVA natural environment. He remains a tireless advocate for this area's wines. Jerry Lohr of J. Lohr Vineyards and Winery has long understood the importance of vineyard soil resource inventory and sustainable vineyard soil management. He is a true agriculturist and an outstanding businessman. I have enjoyed several soils tours within his Paso Robles vineyards. Larry Gomez, winemaker at Lockwood Winery and owner of Via Vega Winery, has been a long-time friend, SCUBA diving buddy and outstanding enologist in our region. Stephan and Beatrice Asseo of L'Aventure Winery and Stephan Vineyards recently gave supportive access to my Cal Poly students to produce a soil survey report for their vineyard. The discoveries made there helped us to better understand the diversity and chemistry of soils derived from the Monterey Formation. Thanks to Stephan and his crew for their generous assistance. A note of appreciation to Jim Moody, a trusty pilot and husband of my colleague, Lynn, for several exciting aerial photographic flights over the Paso Robles AVA region. I wish all the best and continuing success to Kris O'Conner, Executive Director, and her Central Coast Vineyard Team for their many positive contributions to the viticulturists and enologists in this region.

Lastly, and very importantly, I wish to thank the many Cal Poly students, staff, and faculty who have worked cooperatively with me on our many vineyard soils projects and in the production of this book. Notably, Keith Patterson, Richard Smart, Craig Stubler and Ron Taskey have conducted local vineyard projects with me and I thank each of them for their unique contributions and talents. Nathanael Sheean was the artist for the illustrations of the vine soil profile and the orographic figure; Rick Treinen produced the Paso AVA watershed map; and Joe Aroner produced the Paso AVA composite general soils map.

Any errors of omission are mine and I apologize in advance to anyone who was not specifically mentioned and deserved such. Thank you to all of the generous folks who helped me to better appreciate the complexities of the natural environment, to enjoy the fruits (and wines) of their labors, and to better understand the many interrelated facets of viticulture, enology and the business of wine marketing.

As stated by Tracy, my coauthor, this book is only the beginning of our Paso Robles AVA story. Future revisions of this book will be forthcoming as our knowledge deepens and as changes occur.

Thomas J. Rice, Ph.D., C.P.S.S.
Paso Robles, CA

CHAPTER ONE: INTRODUCTION

"GOOD WINE IS A NECESSITY OF LIFE FOR ME"
THOMAS JEFFERSON

WET ROOTS

It was not a terribly auspicious beginning for world-class wine country. For brevity's sake, we will only go back say, 25 million years…the Tertiary Period, of most concern to us, when most of the near-suface landmass we are talking about was forming. There are a number of epochs: the Pliocene (2-5 million years ago) was the major California Coastal Ranges mountain building era. The San Andreas Fault was actively fracturing during the Miocene era (5-24 million years ago) and it has been a real inconvenience ever since.

The Pacific Ocean was influencing the climate of the Central Coast 58 million years ago, much in the same way it does today. During the Miocene-Pliocene epoch transition, some 5-8 million years ago, the California current in the Pacific Ocean was so well developed that a coastal fog regime was in place, which benefits wine growers to this day. The wet and dry orographic (mountain rain shadow) patterns, which influence the dichotomy between the farthest west side to the eastern margins of the Paso Robles American Viticulture Area (AVA) had emerged.

Ancient ancestral interior forests included animal and plant species now found in Asia; Dawn Redwood, Ginkgo, and evolved along the ancestral landscape to include grasslands, savannahs, and woodlands, transitioning to mingle with the northern coniferous forests. About two million years ago, there was pronounced seasonality (dry summers, wet winters) that evolved into the current Mediterranean climate. The land continued to rise into the mountains of the Coast Ranges and the Sierra Nevada to the east, and this further increased regional diversity.

The ice ages of the Pleistocene Epoch (from about 10,000 to 2 million years ago) caused cycles of cooling and warming, as well… with regressing and transgressing (lower and higher) fluctuating sea levels along the coast. These warmer, drier interglacial periods, alternating with the wetter, cooler glacial periods contributed significantly to the composition and evolution of native California flora.

During wetter glacial episodes, coastal forests dominated by Monterey pine, Douglas fir and coastal redwoods migrated south to Santa Barbara and the Los Angeles Basin. During drier interglacial periods, chaparral, coastal scrub and oak-grassland savannah expanded north. The drier climates ultimately caused extinction of the indigenous coastal redwood forest, now largely absent along the coastal lands located south of Big Sur.

About 26,000 years ago, the last glaciation cycle, called the Tioga, began in California, and was then ended due to climate warming, beginning about 15,000 years ago and continuing to this day. During the glacial ice melting periods, large inland lakes generated enormous riparian habitats in a droughty landscape largely devoid of large surface water bodies.

Eventually, humans did arrive…about 11,000 years ago. The Chumash and Salinian indigenous peoples, drawn to the area by the exceedingly rich, varied environment, fully exploited the marine, coastal, and river resources. These Native Americans developed a sophisticated, egalitarian and remarkably artistic society. Their artifacts and presence continue to this day, with several thousand mixed race peoples thriving in the Central Coast area.

EARLY HISTORY: THE IMMIGRANTS

More recent European arrivals included the Portuguese, beginning with Juan Rodriguez Cabrillo in the 1540's, followed by the Spanish conquistadors and the Franciscan missionaries. The Mission San Miguel, eight miles north of the city of Paso Robles, was founded in 1797. The padres immediately planted vineyards for the production of sacramental wine and brandy for export. This was the beginning of organized agriculture in the Paso Robles area. After the secularization of the California missions in the 1840's, the vineyards were abandoned until the next wave of European immigrants. Pierre Hippolyte Dallidet bought the mission's vines in the late 1800's, followed by English entrepreneur Henry Ditmas, who started the area's first commercial vineyard, Rancho Saucelito. Ditmas imported Zinfandel and Muscat vines from France and Spain, and these varietals still thrive in today's Paso Robles vineyards.

Almond orchards were very prevalent from the late 1800's; Paso Robles became known as the Almond City. Viticulture spread past the Mission: by the 1850's the Estrella region, east of the Santa Lucia Mountains, had developed a reputation for producing quality wine grapes. Waves of Irish, Italian, Portuguese, Swedish and Swiss immigrants arrived, drawn westward from the Great Plains by the mild climate and abundant wild game.   Huge parcels of ranch land, such as the 5,000-acre Phelan Ranch, were bought for the cattle and dairy industry. These industries became the backbone of the early 20th century Paso Robles economy.

In 1882, an apple farmer from Indiana named Andrew York arrived and found the area western Paso Robles ideal for grape growing. He founded the first commercial winery, named Ascension Winery. Initial plantings on his 240-acre property included Mission,

Alicante Bouchet and Zinfandel varieties. In particular, his Zinfandel grapes gained a reputation, making Paso well known for this varietal, even back then. The York family made and sold their wines locally, and shipped them in barrels, via horse drawn wagons, to the San Joaquin Valley. Steamboats also delivered the wines northward to San Francisco. The York brothers expanded their operation, buying fruit from nearby vineyards to buttress the growing demand. Because of its distinct soils and climate, York Mountain today has its own appellation, one of the smallest in the United States AVA system. York Mountain remains the oldest continuously operating winery in the county.

Immigrant farming families continued to arrive; the Ernst family from Illinois in 1884, was very successful planting dozens of wine grape varieties. In 1886, Gerd Klintworth planted a vineyard in the Geneseo and Linne area and produced the first known white wine in Paso Robles. Zinfandel again made an appearance west of Templeton in 1890, planted by Frenchman Adolf Siot. Cuttings from the Casteel vineyards in the Willow Creek area, planted in 1908, provided stock for vineyards that are still producing to this day. Lorenzo Nerelli bought a vineyard at the base of York Mountain in 1917, the first to be bonded following Prohibition. THE

HONEYMOON PERIOD: 1920-1970'S

The boom period of the 1920's didn't miss Paso, either. Several families, names that are now local legends, appeared; Dusi, Martinelli, Bianchi and Vosti...and remain here to this day, still being run by families in their third and fourth generations of viticulture. Frank Pesenti planted dryland Zinfandel in 1923; the winery was bonded in 1934. Turley owns the vineyards now...and its Zinfandels legitimately command very high prices. Zinfandel had become a Paso Robles benchmark varietal, and remains one key varietal for the entire appellation. By the 1920's the success of the Paso Robles vineyards caught the eye of a well-known Polish diplomat (a pianist and composer) named Ignace Paderewski. At his Rancho San Ignacio vineyard near Adelaida, he planted the first Petite Sirah in Paso Robles, and grew Zinfandel as well.

By the late 1940's over a dozen small family-owned wineries were planted on both the east and west sides of the Salinas River. During the following decades, development of large-scale vineyards loped along, fueled by the sale of now defunct cattle ranches. The rise in the value of this land, and the correspondingly high taxes, created a need for the California Williamson ("Agricultural Preserve") Act in 1965. This act enables local governments to contract with private landowners to restrict specific parcels of land to agricultural or related open space uses, with tax breaks to agriculturists willing to sign agricultural preserve contracts. Landowners receive significantly reduced property tax assessments. Based on farming and open space uses instead of full market value, this change in the tax law has enabled many agriculturists to keep their properties in productive agricultural enterprises.

In 1964, Dr. Stanley Hoffman founded a vineyard along Peachy Canyon Road in the Adelaida Hills region west of Paso Robles. Next to the old Paderewski ranch, he planted Pinot Noir, Chardonnay and Cabernet Sauvignon, assisted by UC Davis and California's legendary enologist, Andre Tchelistcheff. This was the beginning of the modern era in Paso Robles wine history. University scientific know-how and industrial financing were beginning to affect the region.

The Hoffman Mountain Ranch Winery was the first commercial scale winery in Paso Robles. His Pinot Noir and Cabernet Sauvignon were widely recognized in international wine circles. Other viticulturalists planted the first large plantings on the east side of the Salinas River; Bob Young planted Rancho Dos Amigos on Shandon Heights, Rancho Tierra Rejada, and, in 1973, a 500-acre parcel was planted by Herman Schwartz. Gary Eberle and Cliff Giaocobine planted 700 acres, including the first Syrah in California. They founded Estrella Winery, the largest winery in the area at the time, and have since sold it to a multinational conglomerate.

AVA STATUS AND THE "BIG BOYS:" 1980'S

AVA status was awarded to about 614,000 acres surrounding Paso Robles and the 6,400-acre York Mountain AVA in 1983, encompassing the largest wine growing area in San Luis Obispo County. Large corporate wineries, lured to the area by the high quality of grapes possible with higher yields, an unusual and profitable combination, began to plant large tracts of land. Meridian Winery, also now owned by a multi-national firm, was established in 1988. They now own 3,500 acres, and are producing over a million cases of wine per year, the largest Paso AVA winery. Jerry Lohr expanded his vineyard and winery business to Paso Robles, and owns over 1,900 acres, producing over 450,000 cases annually. Wild Horse, yet another example of winery consolidation, produces 140,000 cases per year in over a dozen different varietals. Mid-sized wineries also flourished during the 1980's: EOS, founded by the Arciero family, with over 700 acres and 160,000 case production, and Treana, founded by the Hope family, with 160,000 cases, thrived as well. It looked like Paso Robles was well on its way to becoming "the breadbasket" of the wine industry, and some interesting changes were beginning to bubble and squeak.

OVERSEAS ATTENTION, RHONE RANGERS, AND WHICH WAY TO MY BED AND BREAKFAST?: 1990'S TO THE PRESENT

Gary Eberle had long since planted Syrah (in the mid-1970's) but Rhone varietals hadn't taken off in Paso Robles until 1989, with the arrival of the Perrin Family. They came to the USA looking for calcareous and limestone-derived soils and a climate similar to their Chateau de Beaucastel estate in the Rhone region of France, where they have produced world class, fine Chateauneuf de Pape wines for generations. Along with their visionary partner, importer Robert Haas, they planted 80 acres in the Adelaida Hills to Rhone varietals (Syrah, Grenache, Roussanne, Marsanne, and Viognier, among others) sourcing exclusive clonal material from their home vineyards in France. They also have made the plant materials for these vines available, via their vine nursery, for sale to other growers. Tablas Creek, as the winery is called, has become one of the leaders for quality Rhone varietal wines in Paso Robles. The region has witnessed a growth explosion for all things Rhone, with less than 100 acres in 1994, to more than 2,000 in 2005. The trend shows no signs of diminishing. The annual Paso Robles Hospice du Rhone is attended by over 3,000 wine aficionados, including a global laundry list of who's-who of Rhone wine grape producers.

There are now over 170 wineries in the Paso Robles Wine Country. The lack of restrictions on a winery's ability to grow, make, blend and market their wines has drawn European winemakers from all over the European Union (EU) seeking relief from restrictive winemaking covenants. Many of these young wineries are establishing strong reputations by producing proprietary Bordeaux, Rhone and Zinfandel blends that the wine world hasn't seen before. The fast growing hospitality segment has mushroomed as well, with a number of larger wineries sporting luxury accommodations, and the smaller ones offering a multitude of reasons to visit...wine caves, horse ranches, olive groves with luxury olive oil production facilities, spa experiences, Native American themed park-like settings...a plethora of shopping and lifestyle opportunities at all levels. The phenomenon of the corporate refugee families donning vigneron's ways continues unabated.

This book will be outdated the moment it is printed, and should serve as a snapshot into the quickly growing and evolving Paso Robles Wine Country. Future editions of this book will incorporate the rapid changes occurring in the dynamic Paso Robles AVA.

A WORKING DEFINITION OF *TERROIR*

Many words have multiple definitions. If you examine a dictionary, you will find that some words have seven or more meanings. Each definition has its unique place when using a specific word in conversation, in writing, or in just thinking about it. The word *terroir*, (pronounced "tair-wah"), is no exception. Indeed, you will rarely find it in English dictionaries since it is a French term.

For every word, it is helpful to understand its evolution in order to better utilize it appropriately. The early roots of *terroir* originate from the Latin word, *terra*, literally meaning earth or land. Many other words also have this same root; e.g., terrace, terrain, terrarium, terrestrial, territory, terror and terrorist. It's curious that a word first used to define earth's land surface has evolved to define persons who defend land from invasion and even strive to acquire new lands by force. Such are the natures of our human languages.

Therefore, it is not surprising to find that *terroir* has many different definitions depending upon the context in which it is used. Many learned scholars have written about *terroir* and its multiple definitions. Today, viticulturists, enologists and marketers use the term to define the entire environment of their grape growing regions and to apply it to the unique regional flavors of their wines. *Terroir* ultimately is used by most persons to express a "sense of place." The *Terroir*-France web site asserts that "a *terroir* is a group of vineyards (or even vines) from the same region, belonging to a specific appellation, and sharing the same type of soil, weather conditions, grapes and winemaking savoir-faire, which contribute to give its specific personality to the wine." Moran (2006) has provided us a comprehensive examination of the many facets of *terroir*. He considers the human influence on *terroir* from the cultivation of land, to selection of plant materials, to the many choices made by the winemaker, from barrel selection to the aging time of the wine, and even to the marketing of wine. If these many facets of *terroir* are to be evaluated and quantified in a scientific context, the variables become too numerous. Therefore, the only unbiased, reasonable methods for *terroir* comparisons among vineyards or wine regions are "blind" wine tastings.

In well designed scientific studies, researchers attempt to control most variables and then allow change of only one or two variables. Therefore, in an attempt to simplify the *terroir* definition so that it is manageable in a scientific context, most researchers will measure the interrelated, environmental variables of geology, soils, landform and climate and their cumulative effects on grapes and wine quality. Even using this definition, there are many independent and dependent variables. Comprehensive measurements of the many variables are often cumbersome and tedious. Likewise, statistical comparisons of the collected data often becomes clumsy and invalid due to the numerous dependent variables.

In this book (and at the risk of trying to define an undefinable concept), our working definition for *terroir* is "the measurable ecosystem variables of a vineyard or a geographic region, where a community of organisms (wine grapevines, their human managers and related biota) interacts with the earth's natural environment in the production of wine grapes."

...our working definition for terroir is "the measurable ecosystem variables of a vineyard or a geographic region, where a community of organisms (wine grapevines, their human managers and related biota) interacts with the earth's natural environment in the production of wine grapes."

"WINE IS BOTTLED POETRY."
ROBERT LOUIS STEVENSON

SOILS OF THE PASO ROBLES AMERICAN VITICULTURAL AREA

SUMMARY

Soils within the Paso Robles American Viticultural Area (AVA) vary regionally and within short distances due primarily to differences in geology (parent material), climatic conditions (precipitation and temperature), landform positions (slope aspect, elevation, and slope steepness), past cropping and other land use histories, and precedent natural biological communities (Figure 1). Additionally, due to the use of mechanical mixing equipment and application of soil amendments, humans influence soil formation and may irreversibly modify soil. Most vineyards in the Paso Robles AVA contain several different soil types, which differ over short distances.

Soils planted to vineyards west of the Salinas River are mainly derived from sedimentary rocks of the Miocene-age Atascadero, Monterey and Santa Margarita Formations and the residual, colluvial and alluvial soil parent materials from these rocks. These rocks are both calcareous (carbonate-rich) and siliceous (silica-rich). Soils derived from the calcareous rocks have very high calcium levels, relatively low potassium and magnesium levels, and alkaline pH's (7.5-8.2) in their subsoils. Soils derived from the siliceous rocks have medium levels of calcium, relatively low potassium and magnesium levels, and acid to neutral pH's (5.5-7.0) in their subsoils. Soil textures are mainly loam, clay loam, silty clay loam and clay, with variable amounts of coarse rock fragments. Within the western Salinas River terraces, the soils form in deep alluvial deposits, which originate from a mixture of rock types.

Soils planted to vineyards east of the Salinas River are also derived from variable parent materials. Soils adjacent to the major creek and river systems are mainly derived from weathered alluvial sediments of the Pleistocene-age Paso Robles Formation along with recent alluvial deposits. These soils have highly variable textures with depth, consisting of stratified layers of clay, gravel and sand. Compacted subsoil horizons often restrict downward internal water flow and may limit vine root growth. Soils derived from the Paso Robles Formation have variable calcium levels, depending on the alluvial parent material sources, have relatively low nitrogen and phosphorus levels, and usually have acid to neutral pH's (5.5-7.0) in their subsoils. There are also many upland hillside soils on the east side of the Salinas River, which are derived from the forementioned geologic formations and resemble the soils located west side of the river.

The following summary table shows the soil distributions on major landforms within the Paso Robles AVA.

Soils on Major Landforms	Acres in Paso Robles AVA	% of Paso Robles AVA
(Total Acres)	614,000	100.0%
Soils on Hills and Mountains	443,418	72.2%
Soils on Terraces	103,211	16.8%
Soils on Alluvial Plains, Alluvial Fans, and Floodplains	67,371	11.0%

SOIL BELOW THE VINE

1. SURFACE ROCK FRAGMENTS

Rock fragments, if present on the soil surface, will retain heat within the topsoil. They also reflect sunlight and warmth back onto the vines, increasing microclimate degree-days and providing frost protection.

2. TOPSOIL (A HORIZON)

This is the main root zone, and is about 15 to 30 cm (6 to 12 in) thick in most soils. Topsoil is formed from a mixture of weathered minerals, organic matter from decaying plants and animals, and fertilizer nutrients.

3. SUBSOIL (B HORIZON)

This is the subsoil zone where leaching products such as clay, carbonates and salts accumulate, and is the foundation for tap roots that physically stabilize the vine and fibrous roots that chemically supply nutrients to the vine via the roots.

4. PARENT MATERIAL (C HORIZON) or BEDROCK (R)

This is the soil parent material and/or underlying bedrock. Residual soils are formed from the weathered rock below, which has direct influences on the soil characteristics. The bedrock is penetrated by roots if it is porous and fractured. Transported parent materials reflect the properties of a mixture of rock types.

5. WATER TABLE

The water table is the top of the saturated soil zone. If groundwater is close to the surface, it will serve as a water source for the vine, provided it is not permanent and does not ultimately limit root growth.

(Illustration by Nathanael Sheean under direction of T.J. Rice)

Figure 1: General Soil Map for the Paso Robles AVA (STATSGO data; USDA-NRCS, 1983; 2003).

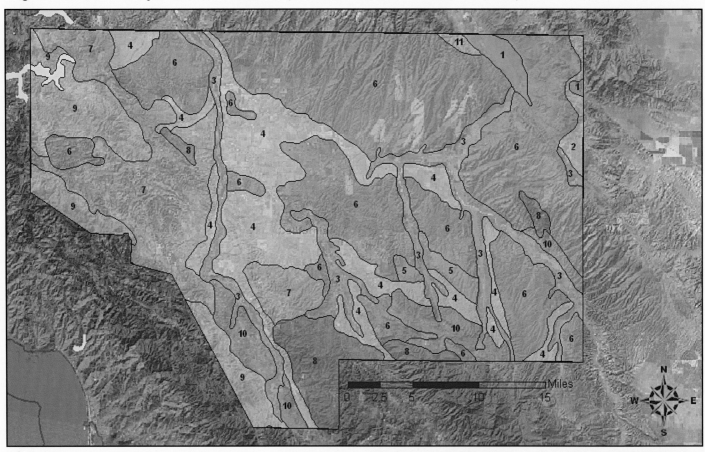

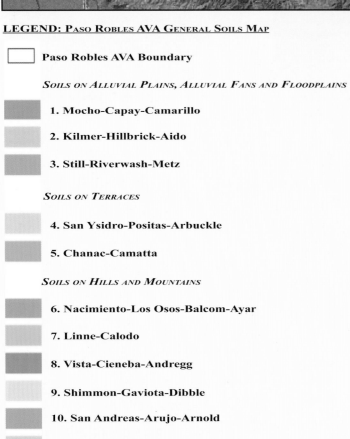

LEGEND: PASO ROBLES AVA GENERAL SOILS MAP

▢ Paso Robles AVA Boundary

SOILS ON ALLUVIAL PLAINS, ALLUVIAL FANS AND FLOODPLAINS

1. Mocho-Capay-Camarillo

2. Kilmer-Hillbrick-Aido

3. Still-Riverwash-Metz

SOILS ON TERRACES

4. San Ysidro-Positas-Arbuckle

5. Chanac-Camatta

SOILS ON HILLS AND MOUNTAINS

6. Nacimiento-Los Osos-Balcom-Ayar

7. Linne-Calodo

8. Vista-Cieneba-Andregg

9. Shimmon-Gaviota-Dibble

10. San Andreas-Arujo-Arnold

11. Shedd-Nacimiento-Los Osos

Source: STATSGO soil data provided by USDA-NRCS (http://soils.usda.gov). Soils data adapted from Soil Survey of San Luis Obispo County, California: Paso Robles Area (USDA, 1983) and Carrizo Plain Area (USDA, 2003).

Map produced by Joe Aroner, Bebeden Wine Co., Paso Robles, CA under the direction of T. J. Rice (August 2007) .

The following soils information is compiled and adapted from the Soil Survey of San Luis Obispo County, California, Paso Robles Area (USDA, 1983) and the Soil Survey of San Luis Obispo County, California, Carrizo Plain Area (USDA, 2003). The "General Soil Map Unit" numbers refer to those that are shown on the general soils map on page 3 (STATSGO; USDA-NRCS, 1983; 2003).

PASO ROBLES AVA   (614,000 ACRES TOTAL; PRWCA, 2007)

A. SOILS ON HILLS AND MOUNTAINS   (443,418 acres; 72.2% of the Paso Robles AVA)

Six of the general soils map units on hills and mountains are within the Paso Robles AVA. The soils are strongly sloping to very steep. Elevation ranges from 600 to 3,400 feet. The soils are shallow to very deep and excessively drained to well drained. They have variable surface layer textures of loamy sand through silty clay.

GENERAL SOIL MAP UNIT 6. NACIMIENTO-LOS OSOS-BALCOM-AYAR SOILS   (254,853 acres; 41.5% of the Paso Robles AVA)

This map unit is located throughout the Paso Robles AVA. Elevation ranges from 600 to 2,500 feet. These soils are moderately deep and deep, strongly sloping to steep, well drained silty clay loams and silty clays. Soils in this unit formed in material weathered from sandstone, mudstone and shale. Slopes range from 9 to 75 percent.

About 40 percent is Nacimiento soils, 15 percent is Los Osos soils, 10 percent is Balcom soils, 10 percent is Ayar soils, and the rest is soils of minor extent. The Nacimiento soils are moderately deep. Typically, they have a surface layer and underlying material of silty clay loam over calcareous sandstone and shale. The Ayar soils are deep. Typically, they have a surface layer of silty clay, and underlying material of clay or silty clay. Typically, the Los Osos soils have a surface layer of clay loam and subsoil of clay underlain by weathered sandstone. Typically, the Balcom soils have a surface layer and a subsoil of loam underlain by weathered, calcareous shale and mudstone. Soils of minor extent are the Calleguas, Arbuckle, Cropley, Diablo, Positas, Rincon and San Ysidro soils. The Cropley and Rincon soils are on alluvial fans. The Calleguas soils are shallow. The Arbuckle, Positas, and San Ysidro soils are on terraces. Xerofluvents and Riverwash are in stream channels. The Diablo soils are deep and have clay textures.

GENERAL SOIL MAP UNIT 7. LINNE-CALODO SOILS   (81,347 acres; 13.2% of the Paso Robles AVA)

This map unit is a large area west of Paso Robles and a smaller area north and east of Atascadero. Elevation ranges from 600 to 1,500 feet. These soils are shallow and moderately deep, strongly sloping to very steep, well drained shaly clay loams and clay loams. The soils in this unit formed in material weathered from calcareous sandstone and shale. Slopes range from 9 to 75 percent (see Figure 2 and Photos 1-4).

About 30 percent is Linne soils, 25 percent is Calodo soils, and the rest is soils of minor extent. The Linne soils are moderately deep. Typically, they have a surface layer and underlying material of shaly clay loam over weathered calcareous shale and mudstone. The Calodo soils are shallow. Typically, they have a surface layer of clay loam over weathered calcareous shale and mudstone.

The soils of minor extent are the Gazos, Balcom, Los Osos, Calleguas, Santa Lucia, Zakme, Diablo, and Lockwood soils. The Gazos and Santa Lucia soils are shaly clay loam. The Balcom soils are calcareous and loamy, and the Calleguas soils are shallow. The Los Osos soils have a clay subsoil. The Zakme and Diablo soils have a clay texture. The Lockwood soils are on terraces (see photo above of backhoe excavating Lockwood soils on Stephan vineyard; 21 August 2006).

GENERAL SOIL MAP UNIT 8. VISTA-CIENEBA-ANDREGG SOILS   (28,498 acres; 4.6% of the Paso Robles AVA)

This map unit is a large area east of Santa Margarita and a small area along Mustard Creek in the east part of the AVA. Elevation ranges from 1,000 to 2,500 feet. These soils are shallow and moderately deep, strongly sloping to very steep, well drained and excessively drained coarse sandy loams. The soils in this unit formed in material weathered from granitic rock. Slopes range from 9 to 75 percent.

About 50 percent is Vista soils, 25 percent is Cieneba soils, 10 percent is Andregg soils, and the rest is soils of minor extent. The Vista soils are moderately deep and well drained. Typically, they have a surface layer and subsoil of coarse sandy loam over granitic rock. They generally are on south slopes of 9 to 50 percent. The Cieneba soils are shallow and excessively drained. Typically, they have a surface layer of coarse sandy loam over granitic rock. Slopes range from 15 to 75 percent. The Andregg soils are moderately deep and well drained. Typically, they have a surface layer and subsoil of coarse sandy loam over granitic rock. They generally are on north slopes of 30 to 75 percent.

The soils of minor extent are Sesame, Hanford, and Metz soils. The sandy loam Sesame soils are on hills, the fine sandy loam Hanford soils are on terraces, and the loamy sand Metz soils are on floodplains.

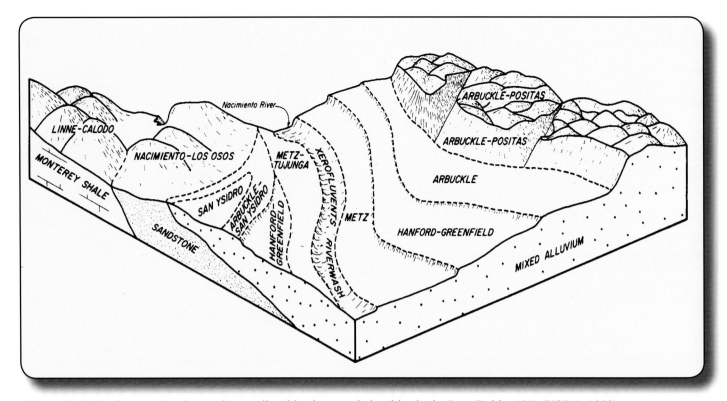

Figure 2: Block diagram showing geology, soil and landscape relationships in the Paso Robles AVA (USDA, 1983).

Photos 1-4: South Viking property (Adelaida Cellars) showing the landscape and soils for Linne shaly clay loam (above) and Calodo shaly clay loam (below). Soils are derived from calcareous mudstone and shale of the Monterey Formation (May 2002).

General Soil Map Unit 9. Shimmon-Gaviota-Dibble Soils (48,418 acres; 7.9% of the Paso Robles AVA)

This area is in the farthest west and southwest of the Paso Robles AVA. Elevation ranges from 1,000 to 2,500 feet. These soils are shallow to moderately deep, strongly sloping to very steep, well drained silty clays, clay loams, and silty clay loams. The soils in this unit formed in colluvial and residual material weathered from sandstone and shale. Slopes range from 9 to 75 percent.

About 25 percent is Shimmon soils, 20 percent is Gaviota soils, 15 percent is Dibble soils and the rest are soils of minor extent. The Shimmon soils are moderately deep. Typically, they have a surface layer of loam and underlying material of clay loam underlain by sandstone. Slopes range from 15 to 75 percent. The Gaviota soils are shallow. Typically, they have a surface layer sandy loam underlain by hard sandstone. Slopes range from 15 to 75 percent. The Dibble soils are moderately deep. Typically, they have a surface layer of clay loam and a subsoil of clay underlain by mudstone and shale. Slopes range from 9 to 75 percent (see photos to the right and below of Dibble soils and associated lands on Stephan vineyard; 24 April 2007).

Soils of minor extent are the Millsholm, Nacimiento, San Andreas, Lompico, Balcom, Arnold, Rincon, and Ryer soils, and areas of rock outcrop. The Millsholm soils are shallow. The Nacimiento soils and Balcom soils are moderately deep, calcareous and loamy. The San Andreas and Lompico soils are on north wooded slopes. The Arnold soils are very deep and sandy. The Rincon and Ryer soils are on alluvial fans. Rock outcrop consists of areas of exposed hard bedrocks.

General Soil Map Unit 10. San Andreas-Arujo-Arnold Soils (28,730 acres; 4.7% of the Paso Robles AVA)

This map unit is composed of small areas around Atascadero, Santa Margarita, and Indian Creek. Elevation ranges from 600 to 2,500 feet. These soils are moderately deep and deep, moderately steep to very steep, well drained and somewhat excessively drained sandy loams, loamy sands, and shaly clay loams. The soils in this unit formed in material weathered from sandstone and shale. Slopes range from 9 to 75 percent.

About 25 percent is San Andreas soils, 20 percent is Arujo soils, 15 percent is Arnold soils, and the rest are soils of minor extent. The San Andreas soils are moderately deep and well drained. Typically, they have a surface layer and subsoil of sandy loam underlain by weathered sandstone. Slopes range from 9 to 75 percent. The Arujo soils are deep and well drained. Typically, they have a surface layer of sandy loam, a subsoil of sandy clay loam and sandy loam underlain by weathered sandstone. Slopes range from 9 to 15 percent. The Arnold soils are deep and somewhat excessively drained. Typically, they have a surface layer of loamy sand and underlying material of sand underlain by weathered sandstone. Slopes range from 9 to 75 percent.

Soils of minor extent are about 10 percent Santa Lucia soils, and 25 percent Lopez, Oceano, Botella, and Concepcion soils. The Santa Lucia soils are moderately deep and well drained. Typically, they have a surface layer of shaly clay loam and very shaly clay loam underlain by hard shale. The Lopez soils are on hills. The Botella soils are on alluvial fans. The Concepcion soils are on terraces, and the Oceano soils are on dunes.

General Soil Map Unit 11. Shedd-Nacimiento-Los Osos Soils   (1,573 acres; 0.026% of the Paso Robles AVA)

This map unit is composed of one small area near the northeast corner of the Paso Robles AVA. Elevation ranges from 600 to 2,500 feet. These soils are moderately deep and deep, strongly sloping to steep, well drained silty clay loams and silty clays. Soils in this unit formed in material weathered from sandstone, mudstone and shale. Slopes range from 9 to 75 percent.

About 40 percent is Shedd soils, 15 percent is Nacimiento soils, 10 percent is Los Osos soils, and the rest is soils of minor extent. The Shedd soils are moderately deep. Typically, they have a surface layer of silty clay loam underlain by soft calcareous shale. The Nacimiento soils are moderately deep. Typically, they have a surface layer and underlying material of silty clay loam over calcareous sandstone and shale. Typically, the Los Osos soils have a surface layer of clay loam and subsoil of clay underlain by weathered sandstone. Soils of minor extent are the Calleguas, Calodo, Cropley, Diablo, Linne, and Rincon soils. The Cropley and Rincon soils are on alluvial fans. The Calleguas and Calodo soils are shallow and underlain by calcareous shale and mudstone. The Linne soils are moderately deep and underlain by calcareous shale and mudstone. The Diablo soils are deep and have clay textures.

B. Soils on Terraces   (103,211 acres; 16.8% of the Paso Robles AVA)

Two of the general soil map units on terraces are within the Paso Robles AVA. The soils are nearly level to very steep. Elevation ranges from 600 to 1,600 feet. The soils in these units are shallow to very deep, well drained, and moderately well drained. They have a surface layer textures of coarse sandy loam to shaly loam.

General Soil Map Unit 4. San Ysidro-Positas-Arbuckle Soils   (96,538 acres; 15.7% of the Paso Robles AVA)

Major areas of this map unit are on terraces adjacent to the Estrella, Huerhuero and Salinas Rivers and their tributaries. Elevation ranges from 600 to 1,500 feet. These soils are very deep, nearly level to hilly, moderately well drained and well drained fine sandy loams, coarse sandy loams, and loams. The soils in this unit formed in alluvium derived from a mixture of rock types on river terraces. Slopes range from 0 to 30 percent (see Figure 3 and Photos 5-8).

About 40 percent is San Ysidro soils, 25 per cent is Positas soils, 15 percent is Arbuckle soils, and the rest is soils of minor extent. The San Ysidro soils are moderately well drained. Typically, they have a surface layer of loam, a subsoil of heavy clay loam, and a substratum of sandy loam. Slopes range from 0 to 9 percent. The Positas soils are well drained. Typically, they have a surface layer of coarse sandy loam, a subsoil of clay, and a substratum of sandy loam. Slopes range from 9 to 30 per cent. The Arbuckle soils are well drained. Typically, they have a surface layer of fine sandy loam, a subsoil of sandy clay loam, and a substratum of sandy loam. Slopes range from 0 to 30 percent.

Soils of minor extent are Hanford, Greenfield, Nacimiento, Mocho, Metz, Tujunga, and Rincon soils, and Xerofluvents and Riverwash. The Hanford and Greenfield soils are on lower alluvial terraces. The Mocho and Rincon soils are on alluvial fans. The Nacimiento soils are on hills and mountains. The Metz and Tujunga soils are on floodplains. The Xerofluvents and Riverwash are in stream channels.

General Soil Map Unit 5. Chanac-Camatta Soils   (6,673 acres; 1.1% of the Paso Robles AVA)

This map unit is east of Shedd Canyon. Elevation is about 1,600 feet. These soils are very deep, gently rolling to very steep, well drained loams; some are shallow to a hardpan. The soils in this unit formed in alluvium derived from mixed rock on high terraces. Slopes range from 5 to 75 percent.

About 45 percent is Chanac soils, 40 percent is Camatta soils, and the rest is a soil of minor extent. The Chanac soils are very deep. Typically, they have a surface layer and subsoil of loam and a substratum of loam and fine sandy loam. Slopes range from 9 to 75 percent. The Camatta soils are shallow to a hardpan. Typically, they have a surface layer of loam overlying an indurated lime cemented hardpan. Below the hardpan is a very fine sandy loam substratum. Slopes range from 5 to 30 percent. The soil of minor extent in this unit is Polonio soil. It is on alluvial fans.

C. Soils on alluvial plains, alluvial fans, and floodplains   (67,371 acres; 11.0% of the Paso Robles AVA)

Three of the general soil map units are on alluvial plains, alluvial fans, and floodplains within the Paso Robles AVA. The soils are adjacent to the major stream channels of the Estrella, Huerhuero, Salinas, and San Juan Creeks and Rivers & their tributaries. Slopes range from nearly level to moderately sloping. Elevation ranges from 600 to 1,500 feet. The soils in these areas are very deep, and poorly drained to somewhat excessively drained. Surface layer textures range from loamy sand to silty clay.

General Soil Map Unit 1. Mocho-Capay-Camarillo Soils (6,845 acres; 1.1% of the Paso Robles AVA)

This map unit is located along Cholame Creek in Cholame Valley. Elevation is about 1,100 feet. These soils are very deep, nearly level to moderately sloping, poorly drained to well drained clay loams, silty clays, and silty clay loams. The soils in this unit formed in alluvium derived from sedimentary rock on alluvial fans and floodplains. Slopes range from 0 to 9 percent.

About 40 percent is Mocho soils (see photo to the right of Mocho sandy loam on a stream terrace), 25 percent is Capay soils, 15 percent is Camarillo soils, and the rest of the map unit is composed of soils of minor extent. The well drained Mocho soils are on alluvial fans and floodplains. Typically, they have a surface layer of clay loam and underlying material of stratified clay loam, loam, or silty clay loam. Slopes range from 0 to 9 percent. The moderately well drained Capay soils are on floodplains. Typically, they have a clay texture surface layer and silty clay in the subsoil. Slopes range from 0 to 2 percent. The poorly drained Ca12marillo soils are on floodplains. Typically, they have a surface layer of silty clay loam and underlying material of stratified fine sandy loam, loam, or silty clay loam. Slopes range from 0 to 2 percent. The soils of minor extent are Clear Lake and San Emigdio soils. Clear Lake soils are in the basins. San Emigdio soils are on alluvial plains and fans.

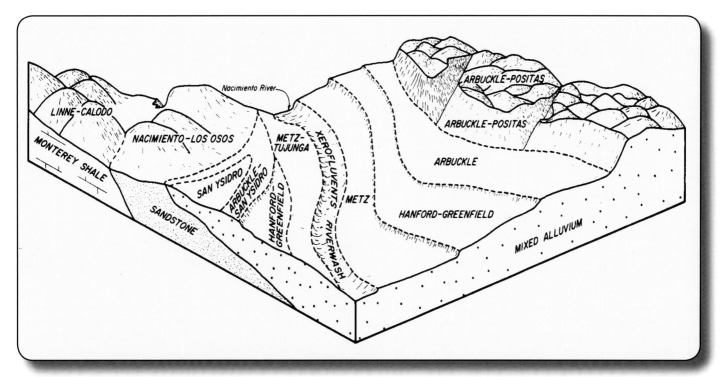

Figure 3: Block diagram showing geology, soil and landscape relationships in the Paso Robles AVA (USDA, 1983).

Photos 5-8: Arbuckle soils formed in Paso Robles Formation on J. Lohr Hilltop Vineyard (above; 7 May 2001) and Arbuckle and Positas soils formed in Paso Robles Formation alluvium along Dry Creek next to the Steinbeck Vineyards (below: 10 Nov. 2006).

This map unit is in the far eastern part of the Paso Robles AVA. Elevation ranges from 200 to 2,500 feet. These soils are shallow to moderately deep, nearly level to steeply sloping, well drained sandy loams, loams and clay. The soils in this unit formed in residuum derived from sedimentary rock on hills and mountains. Slopes range from 5 to 75 percent.

About 30 percent is Kilmer soils, 20 percent is Hillbrick soils, 15 percent is Aido soils, and the rest is soils of minor extent. Typically, Kilmer soils have a surface layer and subsoil of loam and clay loam underlain by massive calcareous shale. Typically, Hillbrick soils have a surface layer of sandy loam underlain by fractured calcareous shale. Typically, Aido soils have a surface layer and subsoil of clay underlain by weathered shale.

Soils of minor extent are Mocho, Still, Metz, Tujunga, Xerofluvents, and Riverwash. Metz and Tujunga soils are on floodplains. Xerofluvents and Riverwash are in stream channels. Mocho and Still soils are on alluvial plains.

GENERAL SOIL MAP UNIT 3. STILL-RIVERWASH-METZ SOILS
(55,139 ACRES; 9.0% OF THE PASO ROBLES AVA)

This map unit is along the Salinas River and Toro Creek, in the San Margarita area and along the San Juan and Estrella Creeks in the Creston and Shandon areas. These soils are very deep, nearly level to moderately sloping, well drained and somewhat excessively drained clay loams, loams, and loamy sands. Elevation ranges from 600 to 1,500 feet. The soils in this unit formed in alluvium derived from mixed rocks on alluvial fans, alluvial plains, and floodplains. Slopes are 0 to 9 percent.

About 30 percent is Still soils, 20 percent is Riverwash, 15 percent is Metz soils, and the rest is soils of minor extent. The well drained Still soils are on alluvial plains and fans. Typically, they have a surface layer and underlying clay loam materials. Slopes range from 0 to 9 percent. Riverwash is on floodplains and alluvial plains and fans. Typically, they have a surface layer of gravelly sandy loam and underlying material of stratified loam, sandy loam, sand or loamy sand. Slopes range from 0 to 9 percent. The somewhat excessively drained Metz soils are on floodplains. Typically, they have a surface layer of loamy sand and underlying material of stratified sand, loamy sand, and very fine sandy loam. Slopes range from 0 to 5 percent. Soils of minor extent are 15 percent Xerofluvents (see the photo to the above right of a gravelly Xerofluvents soil on a stream terrace) and Riverwash in stream channels, and 20 percent Clear Lake, Elder, Hanford, and Mocho soils. The Clear Lake soils are in basins, and the Elder and Hanford soils are on first-level stream terraces. The Mocho soils are on alluvial plains.

Photo 9: Aerial view of Santa Margarita Ranch to the southwest over the Santa Lucia Mountains toward Estero Bay (20 March 2003).

TAXONOMIC CLASSIFICATION OF THE SOILS

The system of soil classification used by the National Cooperative Soil Survey has six categories (Soil Survey Staff, 1999; 2006). Beginning with the broadest, these categories are the order, suborder, great group, subgroup, family, and series. Classification is based on soil properties observed in the field or inferred from those observations or from laboratory measurements. This table shows the classification of the soils in the survey area. The categories are defined in the following paragraphs.

ORDER. Twelve soil orders are recognized on earth. The differences among orders reflect the dominant soil-forming processes and the degree of soil formation. Each order is identified by a word ending in -sol. An example of a soil order is Alfisols.

SUBORDER. Each order is divided into suborders primarily on the basis of properties that influence soil formation and are important to plant growth or properties that reflect the most important variables within the orders. The last syllable in the name of a suborder indicates the order. An example is Xeralfs (Xer-, meaning xeric, plus -alfs, from Alfisols).

GREAT GROUP. Each suborder is divided into great groups on the basis of close similarities in kind, arrangement, and degree of development of pedogenic horizons; soil moisture and temperature regimes; type of saturation; and base status. Each great group is identified by the name of a suborder and by a prefix that indicates a property of the soil. An example is Haploxeralfs (Hapl-, meaning minimal horizonation, plus xeralfs, the suborder of the Alfisols that has a xeric moisture regime).

SUBGROUP. Each great group has a typic subgroup. Other subgroups are intergrades or extragrades. The typic subgroup is the central concept of the great group; it is not necessarily the most extensive. Intergrades are transitions to other orders, suborders, or great groups. Extragrades have some properties that are not representative of the great group but do not indicate transitions to any other taxonomic class. Each subgroup is identified by one or more adjectives preceding the name of the great group. The adjective Typic identifies the subgroup that typifies the great group. An example is Typic Haploxeralfs.

FAMILY. Families are established within a subgroup on the basis of physical and chemical properties and other characteristics that affect management. Generally, the properties are those of horizons below plow depth where there is some biological activity. Among the properties and characteristics considered are particle size class, mineralogy class, cation exchange activity class, soil temperature regime, soil depth, and reaction class. A family name consists of the name of a subgroup preceded by terms that indicate soil properties. An example is fine-loamy, mixed, superactive, thermic Typic Haploxeralfs.

SERIES. The series consists of soils within a family that have horizons similar in color, texture, structure, reaction, consistence, mineral and chemical composition, and arrangement in the profile. There are over 23,000 soil series recognized in the USA. An example from the Paso Robles AVA is the Calodo series.

SOIL ORDERS, SUBORDERS, AND GREAT GROUPS

This section discusses the orders, suborders, and subgroups mapped in the Paso Robles soil survey area (USDA, 1983). Unless otherwise stated, the soils in this area have a *xeric* moisture regime and a *thermic* temperature regime. Therefore, unless the soil is irrigated, the moisture control section is moist in some part from December until May and is dry in all parts from July until October in 6 out of 10 years. The mean annual soil temperature ranges from 59 to 62 degrees F. The Paso Robles airport weather station, where the climatic data in Table 5 was obtained, is in the central part of the area. However, the moisture regime varies from xeric to xeric bordering on aridic, and the temperature regime varies from thermic to mesic in the survey area.

The vineyard soils of the Paso Robles AVA are classified into five (5) soil orders: Alfisols, Entisols, Inceptisols, Mollisols and Vertisols. In the following paragraphs, each of the five soil orders and their categories of suborders and great groups are discussed (Soil Survey Staff, 2006).

ALFISOLS

Alfisols are soils in the area that have a massive and hard A horizon and a finer textured B horizon. They have high basic cation saturation in which water is held at less than 15-bar tension during at least three months each year when the soil is warm enough for plants to grow.

Alfisols in this area have been placed in the Xeralfs suborder, which has a clay-enriched B horizon (see the photo to the right showing Arbuckle and Positas soils, both Alfisols, formed in the Paso Robles Formation; Dry Creek,

10 November 2006). The Xeralfs are divided into two great groups: the Haploxeralfs and Palexeralfs. Soils that have less than 15 percent clay increase between the A and B horizons have been placed in the Haploxeralfs great group. Soils that have an abrupt boundary with more than 15 percent clay increase between the A and B horizons have been placed in the Palexeralfs great group.

ENTISOLS

Entisols are soils in this area that have little or no evidence of development of pedogenic horizons (see the photo to the right and below showing Calleguas soils, a Xerorthent, formed in calcareous mudstone and shale from the Monterey Formation; Stephan Vineyard, 26 April 2007). These Entisols are in the Orthent, Fluvent, and Psamment suborders. All of these soils lack a subsoil B horizon and generally have organic matter of less than one percent.

The Orthents are loamy very fine sand or finer in the textural control section. The organic matter content decreases regularly with depth to less than 0.3 percent at a depth of 50 inches. Most of these soils are on alluvial fans and terraces. A few are on hills and mountains.

Fluvents are similar to Orthents, except that the organic-matter content decreases irregularly to a depth of at least 50 inches. These deep and very deep soils are on alluvial fans and plains where water has deposited stratified soil parent materials that buried topsoils and subsoils of older preexisitng soils.

PASO ROBLES HAPLOXERALFS

Suspended clay and silt particles settled in offshore ocean waters.
Uplifting tectonic forces raised the ancient aquatic sediments
Above sea level and the earth's aerobic atmosphere dried them.

Fracturing and folding of the substratum bedrock developed
Rolling hills and valleys. Eroding rains in wetter Ice Age climates
Washed rock & soil materials off the hills onto the alluvial plains.

Grasses, forbs and brush vegetated the new alluvial
Soil deposits and sent roots deep into the regolith.
Humus formed from the decomposing sinuous roots.

Topsoils formed and gophers and ground squirrels
Burrowed actively; feeding on the luscious roots, and
Homogenized the churned soil and rock deposits.

Today, soft porous fertile topsoils support lush seasonal
Xeric vegetative communities. Rabbits, deer, fowl and
Cougars roam the stream terraces and floodplains.

CALIFORNIA XERORTHENTS

Prehistoric ocean sediments enriched in calcium carbonates were
Uplifted far above present sea levels and formed sedimentary rocks.

Rocks are now subjected to atmospheric weathering processes and
Thin soils form in the softened sediments; vegetation colonizes soils.

Decomposing grasses are consumed by bacteria and fungi;
Animal waste products decompose into humus and organic acids.

Carbonic acids etch the calcareous sediments; ancient fish scales, and
Marine animal skeletons dissolve in the acid bath and release nutrients.

Mediterranean climatic grass & chaparral lands are now examined for
Conversion to new vineyards to be planted with European vine species.

Psamments are loamy fine sand or coarser in the control section. Most of these soils are on fans & terraces. A few are on hills and mountains. The Orthents, Fluvents, and Psamments have been placed in the Xerorthent, Xerofluvent, and Xeropsamment great groups because they have a xeric moisture regime. Soils with a xeric moisture regime bordering on an aridic moisture regime have been placed in the Torriorthent great group.

INCEPTISOLS

Inceptisols are soils in this area that have slightly weathered horizons that have lost basic cations (such as Ca, K, Mg, or Na) or iron (Fe) and aluminum (Al) but retain some weatherable minerals.

The Inceptisols in this area are in the Ochrepts suborder. They have an ochric epipedon and a cambic horizon. The color, organic matter, or structure is lacking for a mollic epipedon. The cambic horizon has a one to two percent clay increase and has soil structure. The texture is coarse sandy loam or finer. These soils are on hills and mountains. Because they have a xeric moisture regime, they are placed in the Xerochrepts (pr., *zear-o-krepts*) great group.

Mollisols

Mollisols are grassland soils that typically have a dark surface layer more than 10 inches thick that has more than one percent organic matter and is not both hard and massive (see the photo below of Danville soil, an Argixeroll; HMR Vineyard area, 3 May 2001). Most of these soils have a thermic temperature regime. Lompico, McMullin, and Zakme soils, however, have a mesic temperature regime.

CALIFORNIA ARGIXEROLLS

Ancient muddy ocean sediments enriched in calcium carbonates were
Deposited in deep marine environs; these were formerly the burial
Grounds for multitudes of fishes and microscopic planktonic organisms.

Earthly tectonic movements of the North American and Pacific Plates
Caused uplift of the sediments; building the California Coast Ranges.

Dried sedimentary rocks are folded and faulted; subjected to atmospheric weathering
Processes, soils form in the decomposing sediments; plants colonize the soils.

Decomposing grass roots are consumed by bacteria and burrowing animals;
Animal waste products decompose into humus and organic carbonic acids.

Mineral particles are coated with the brown organic humus and deep
Topsoils, rich in plant and animal remains provide nourishment to the vegetation.

Carbonic acids decompose calcareous sediments; calcium carbonate clays dissolve and
Leach deeper into the soils; precipitate as white secondary carbonates in subsoil zones.

California coastal Mediterranean climates produce winter seasonal rains which
Infiltrate the porous topsoils and percolate into the underlying shales and mudstones.

Soils resemble those in the historic vineyards of the Rhone terraces of southern France,
Bordeaux vinelands of eastern France, and Tuscany-Chianti wine regions of Italy.

The Mollisols in this area are in the Xeroll and Alboll suborders. The Xerolls and Albolls are divided into three great groups: Haploxerolls, Argixerolls, and Argialbolls. Soils that do not have a clay-enriched B horizon and generally lack layers strong in calcium carbonate are classified in the Haploxeroll great group. Soils that have a clay-enriched B horizon, a clear to gradual boundary between the A and B horizons, and lack strong calcium carbonate layers are placed in the Argixeroll great group. Soils that have an albic horizon that lies immediately below the dark surface layer and have a clay-enriched B horizon are classified as Argialbolls.

Vertisols

Vertisols are soils that are fine textured throughout and mainly consist of montmorillonite clays (>30%) that swell and shrink significantly on wetting and drying (see photo below of cracked surface clays in a Vertisol). Unless irrigated, these soils dry in summer and crack from the surface downward to a depth of at least 20 inches. Because of negligible rainfall in the summer, these cracks remain open for more than 90 consecutive days each year.

Typically, these soils have an A horizon that is firm and massive when moist, but becomes granular or blocky and hard or very hard when dry. The surface soil falls into the cracks and causes internal displacement. Because of the internal churning that takes place in these soils, development of a B horizon is not possible. This churning also results in the formation of intersecting, shiny slickenside faces in the clayey substratum. Vertisols in this area have been placed in the Xerert suborder, which has a xeric moisture regime. All of the Xererts are Haploxererts (pr., *hap-lo-zear-erts*) (see lower photos on page 16 of Cropley soils, a Haploxerert; Halter Ranch Vineyard, 18 August 2002).

CALIFORNIA ADELAIDA ZAKME SOIL

Rich fertile organo-mineral topsoil
Covers the alkaline calcareous substratum.

Fibrous grass roots have decayed and added
Humus and nutrients to the topsoils for centuries.

Carbonic and humic acids with rainfall have decomposed
Calcium carbonates and leached them into subsoil layers.

The resultant soil sustains wine grape vineyards
Throughout the earth's great temperate viticultural areas.

Bordeaux and Rhone in France, Tuscany in Italy,
Adelaide in Australia, and Adelaida in California.

Analogues among regions with Mediterranean climates;
Marine deposited sedimentary rocks high in carbonates.

A renaissance of appreciation for French *terroir*; where
Cabernet Sauvignon, Cabernet Franc, Grenache, Merlot,
Mourvedre, Pinot Noir and Syrah scions flourish.

PASO ROBLES HAPLOXERERTS

Tectonic plates covered with ocean sediments and underlain by
Deep crustal ultramafic rocks are pasted onto the continent's edge.
Upheaval and resulting earthquake tremors rattle the earth,
Lifting these lithologies onto the dry upland surfaces.

Weathering and decomposition produces vermiculite and
Montmorillonite clays that erode onto valley floors.
Inverted soils develop within these marine parent materials.
Organic matter and grass humus darkens the surface and subsoil.

Mediterranean climates wet and dry the soils each year and
Deep subsoil cracks swallow and engulf organic rich topsoils.
Animals step gingerly on the dry, cracked smectite clays to
Avoid falling into them and suffering broken limbs.

Buildings and roads constructed directly on these soils will
Ultimately crack and crumble in short periods of time.
Saturated clayey hillslopes are subject to mass movements as
Rotational slumps and complex blocks of land succumb to gravity.

The photos to the right illustrate a soil profile of Zakme soils (clayey Mollisols) within a Cabernet Sauvignon vineyard landscape at Halter Ranch Vineyard. These Zakme soils form in montmorillonite clay (expansive cracking clays), alluvial parent materials over calcareous shale and mudstone of the Monterey Formation, Sandholdt member. These soils are located on lower valley positions within many vineyards in the Adelaida Hills region, including Carmody McKnight (photo below), Halter Ranch, Justin, and Tablas Creek.

Table 1: Major Vineyard Soils' Classifications in the Paso Robles AVA

Soil Series	Soil Order	Soil Great Group
Arbuckle	Alfisols	Haploxeralfs
Ayar	Vertisols	Haploxererts
Balcom	Inceptisols	Xerochrepts
Botella	Mollisols	Argixerolls
Calleguas	Entisols	Xerorthents
Calodo	Mollisols	Haploxerolls
Chanac	Inceptisols	Xerochrepts
Cropley	Vertisols	Haploxererts
Diablo	Vertisols	Haploxererts
Dibble	Alfisols	Haploxeralfs
Elder	Mollisols	Haploxerolls
Gaviota	Entisols	Xerorthents
Gazos	Mollisols	Haploxerolls
Greenfield	Alfisols	Haploxeralfs
Linne	Mollisols	Haploxerolls
Lockwood	Mollisols	Argixerolls
Lodo	Mollisols	Haploxerolls
Lopez	Mollisols	Haploxerolls
Los Osos	Mollisols	Argixerolls
Millsholm	Inceptisols	Xerochrepts
Mocho	Mollisols	Haploxerolls
Nacimiento	Mollisols	Haploxerolls
Pico	Mollisols	Haploxerolls
Positas	Alfisols	Palexeralfs
San Andreas	Mollisols	Haploxerolls
San Ysidro	Alfisols	Palexeralfs
Santa Lucia	Mollisols	Haploxerolls
Sorrento	Mollisols	Haploxerolls
Still	Mollisols	Haploxerolls
Zakme	Mollisols	Haploxerolls

Table 2: Soils derived from calcareous soil parent materials (subsoil pH>7.0; alkaline soils).

Soil Series	Soil Depth	Soil Vigor Potential
Ayar	Very Deep	High
Balcom	Deep	High
Botella	Very Deep	High
Calleguas	Shallow	Low
Calodo	Shallow	Low
Diablo	Deep	High
Linne	Moderately Deep	Moderate
Nacimiento	Moderately Deep	Moderate
Sorrento	Very Deep	High
Zakme	Deep	High

Table 3: Soils derived from siliceous or non-calcareous parent materials (subsoil pH<7.0; acid soils).

Soil Series	Soil Depth	Soil Vigor Potential
Gaviota	Shallow	Low
Gazos	Moderately Deep	Low
Lockwood	Very Deep	High
Lodo	Shallow	Low
Lopez	Shallow	Low
Los Osos	Moderately Deep	Moderate
San Andreas	Moderately Deep	Low
Santa Lucia	Deep	Moderate

Table 4: Soils derived from mixed alluvium parent materials (variable soil pH).

Soil Series	Soil Depth	Soil Vigor Potential
Arbuckle	Very Deep	High
Camarillo	Very Deep	High
Cropley	Very Deep	High
Dibble	Very Deep	High
Elder	Very Deep	Moderate
Greenfield	Very Deep	Moderate
Mocho	Very Deep	High
Positas	Very Deep	High
San Ysidro	Very Deep	High

FACTORS OF SOIL FORMATION & RELATED VITICULTURAL IMPLICATIONS OF THE PASO ROBLES AVA

This section discusses the factors of soil formation, relates them to the formation of soils in the survey area, and explains the processes of soil formation.

Soil, in which plants grow, is a natural body on the surface of the earth. It is a mixture of rocks and minerals, organic matter, water, and air, all of which occur in varying proportions. The rocks and minerals are usually weathered and fragmented. Soils have distinctive layers, called horizons, that are the products of earth's environmental forces acting upon materials that are deposited or weathered in place by geological and atmospheric processes.

The characteristics of a soil are determined by the interaction of: (1) the chemical, physical and mineralogical composition of the parent material; (2) the climate in which the soil material has accumulated and has existed since accumulation; (3) the relief, or topography, which influences the local, or internal, environment of the soil, its drainage, moisture content, aeration, susceptibility to erosion, and exposure to sun and wind; (4) biological forces that act upon the soil material, such as the plants and animals living on and in the soil; and (5) the length of time the forces of development have acted on the soil material.

These five soil-forming factors are dependent on each other. In the following sections are discussions of how each soil-forming factor affects the soils within the Paso Robles AVA.

SOIL PARENT MATERIAL

In this section, some of the most extensive soils are mentioned in conjunction with their parent materials. Many soil types are mapped on more than one geologic formation, because different formations, or varying components of a formation, give rise to similar soils.

The largest single geologic unit in the area is the Paso Robles Formation. It is north of the La Panza Range, east of the Nacimiento Fault zone, and widely distributed along the Santa Margarita syncline. In many places, it overlies the Monterey Formation, Vaqueros Formation, Santa Margarita Formation and granitic rocks.

Parent materials, in which the soils in the Paso Robles area have developed, are residual, colluvial and alluvial. The Paso Robles Formation consists of poorly sorted gravel, sand, silt, and clay. The composition varies, depending upon the original rock sources. In many areas it is difficult to distinguish between older alluvium and this formation. In many areas the Nacimiento and Los Osos soils are the main soils, with terrace cappings of Arbuckle, Positas and San Ysidro soils.

The exposed Cretaceous granitic rocks mainly occur east of Santa Margarita. The rocks are deeply weathered and commonly decomposed, resulting in a highly eroded terrain. Minerals in this rock mainly consists of biotite and adamellite. Coarse-textured soils are derived from granite. The Andregg, Cieneba and Vista soils are the main soils on this material. However, Sesame soil, with a moderately fine textured subsoil, formed on many lower slopes.

The Monterey Formation is a well-bedded marine Miocene sequence mainly composed of siliceous and calcareous sedimentary rocks. The Gazos, Lockwood, Lopez, McMullin and Santa Lucia soils are mainly on siliceous shale, and the Calodo, Linne, Nacimiento and Zakme soils are mainly derived from calcareous shale and mudstone (see the photos on page 14 showing the landscape and profile of Zakme soils, clayey Mollisols weathered from calcareous shale; taken on a footslope located at Halter Ranch, 16 August 2002).

The Dibble, Gaviota, and Shimmon soils are the most extensive soils mapped on the Atascadero Formation and the sandstone and conglomerate from an unnamed formation. These formations are from the upper Cretaceous Period. The Atascadero Formation has interbedded layers of siltstone and mudstone of varying thickness. The complexity of the bedding gives rise to a wide range of soil textures and soil depths. Moderately deep, fine textured Dibble soils formed on the Atascadero Formation. Shallow, coarse textured Gaviota soils formed on an unnamed sandstone and conglomerate formation.

The Santa Margarita Formation consists of thick beds of weakly consolidated arkosic sandstone. This formation is mainly exposed around Atascadero and Santa Margarita. The Concepcion soil, which is the most developed soil in this area, is found only on this formation. The San Andreas and Arnold soils are the main soils, but these soils also occur on other formations.

Because many of the streams dissect different geologic formations, most of the alluvial material is heterogeneous and highly mixed. The composition of gravel, sand, silt, and clay of the alluvium has been altered during water movement. The coarse textured, sandy soils are found mainly at the beginning of streams or where the stream velocity is high. The Metz and Tujunga soils are examples of these coarse textured soils, and are found adjacent to or in the present stream channels. The fine textured (more clay) soils are found on stream terraces where the water velocity is low or in basin locations. The Clear Lake and Camarillo soils are examples of soils that formed in areas where water remained long enough to allow the clays settle.

As the regional land surface has uplifted due to tectonic forces, many terraces have formed along the stream channels (see Figure 4 and Photos 10-13). The age of the older alluvium is generally determined by the degree of horizon development in the soil profile. Geologically, the older alluvium is mainly of Pleistocene age. Similarity of the composition between the older alluvium and the modern stream channel deposits indicates that drainage patterns have remained fairly constant. However, the parent material of the Clear Lake soil and other geographically related soils suggests that there may have been a different drainage pattern in the area where this terrace material was deposited.

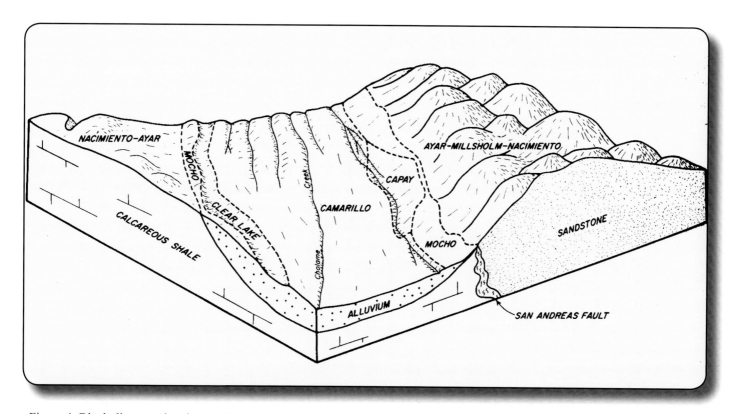

Figure 4: Block diagram showing geology, soil and landscape relationships in the Paso Robles AVA (USDA, 1983).

Photos 10-13: HMR Vineyard west (Adelaida Cellars) showing the vineyard landscape and soils for Nacimiento clay loam (above) and Balcom clay loam (below). Soils are derived from calcareous mudstone and shale of the Monterey Formation (April 2001).

CLIMATE AND SOIL FORMATION

The climate, or the amount and distribution of solar radiation (heat) and moisture received, has a significant influence on the kind of soil that forms. Heat and moisture strongly influence the amounts and types of vegetation, the rate at which organic matter decomposes, the rate at which the minerals weather, and the erosion or accumulation of soil horizons.

Most of the climate in the Paso Robles area is a subhumid, mesothermal climate that is characterized by cool, moist winters and hot, dry summers (see Table 5 and Figures 6-7).

Differences in rainfall and temperature are influenced by the elevated and mountainous topography on the western edge of the survey area. Mean annual precipitation ranges from over 30 inches in the western part of the area to nine (9) inches in some areas of the eastern part. Mean annual air temperature ranges from about 56 degrees F in the western part of the area to 60 degrees F throughout the rest of the area.

Effects of higher precipitation and lower temperatures are evident in the lush vegetation and the soils, which form in the Santa Lucia Range. Woody and herbaceous vegetation is more abundant and the soil organic matter levels are higher. Also, the higher soil acidity (lower pH) and lower soil base cation (less Ca, K, Mg, and Na) saturation indicate greater soil leaching by infiltrating waters. The Lompico and McMullin soils are examples of soils occurring in this mountainous area.

As the temperature increases and rainfall decreases, woody tree vegetation is generally restricted to the north slopes. In contrast, on the south slopes there are lighter colored, massive soils. Millsholm soils are examples of soils on south aspects, while Dibble soils are located on north aspects The Camatta and Polonio soils are examples of soils in the areas of lowest rainfall and highest temperatures in the eastern area of the Paso Robles AVA.

PASO ROBLES AVA CLIMATE (adapted from Elliott-Fisk, 2007)

Geographically, the Paso Robles area is characterized by a maritime climate, with smaller monthly temperature ranges than areas further inland. This climate has long been recognized by botanists and agronomists as relevant to both the native vegetation of the region and the regional potential for agriculture. Local citizens experience marine air spilling across the crest of the Santa Lucia Range from the cool Pacific Ocean and feel the sea breezes that accompany this air mass. Occasionally, sea breezes from Monterey Bay far to the north also migrate southward through the Salinas River Valley, reaching San Miguel and Paso Robles. The frequency and duration of sea breezes diminish to the east, especially east of the proposed Creston District and Paso Robles Estrella District areas.

The climate of the Paso Robles AVA using the global scale climate classification system of Koppen, Geiger and Pohl (1953) is Mediterranean warm summer (Csb), but a small northeast portion of the Paso Robles AVA is warmer as a Mediterranean hot summer climate (Csa). However, long-term climate data comparisons for the cities of Paso Robles (inside of the Paso Robles AVA), for the city of San Luis Obispo to the southwest, and for the city of Fresno to the east in the Central Valley of California, demonstrate Paso Robles' climate to be more maritime than continental in nature.

Using the background of this regional climate, smaller scale local climates with the Paso Robles AVA are found along gradients of:

1. **Longitude**: continentality, with areas further east more continental;

2. **Elevation**: adiabatic and orographic influences on temperature, dewpoint, and precipitation following elevational gradients;

3. **Proximity to the ocean**: maritime influence, increasing to the west towards the Pacific Ocean and below specific topographic gaps in the range crest; and

4. **Topography**: mountain-valley position and location along the major rivers and creeks influencing wind flow, the incursion of marine air masses, and setting up local winds with surface heating and cooling.

The variations in temperature, precipitation, evapotranspiration, wind, cloud and fog cover and its duration, growing degree days, and other climate variables also are of great significance for all plant life (natural vegetation and agronomic crops, including winegrapes), influencing phenological events in the vine's life cycle, canopy development, fruit set and maturation, and osmotic stress via transpiration. Taking these various factors into account, distinct climatic gradients exist across the large Paso Robles AVA, with the following:

1. Annual precipitation is highest to the southwest, in the Santa Margarita Ranch area, due to proximity to the coast and higher elevation, decreasing to both the north towards the area of the San Miguel District, and to the east towards the town of Shandon and the San Juan Creek area;

2. Temperatures are mildest, or most maritime, to the west and on the slopes of the Santa Lucia Range, including the areas of the proposed Templeton Gap, Paso Robles Willow Creek District, Adelaida District, and Santa Margarita Ranch viticultural areas; more moderate conditions exist in the city of Paso Robles and the areas of the proposed El Pomar District, Creston District, Geneseo District and the Paso Robles Estrella District viticultural areas, with summer and fall incursions of marine air masses common on cooler days. Warmer conditions exist in the areas of the proposed San Miguel District, San Juan Creek and Paso Robles Canyon Ranch viticultural areas, although cold air drainage in the evenings moves east down the mountain slopes and reduces evening temperatures and, thus, degree-day totals. Therefore, both degree-day totals and temperature ranges are greatest in the areas of the proposed San Juan Creek

and Paso Robles Canyon Ranch viticultural areas, and lowest in the areas of the proposed Templeton Gap, Paso Robles Willow Creek District, Adelaida District and Santa Margarita Ranch viticultural districts.

3. Other climatic parameters such as wind, evapotranspiration, fog and cloud cover are important for the grapevines, but long-term data for such parameters are rarely available.

The southwestern part of the Paso Robles AVA is along the crest and eastern slope of the Santa Lucia Range, one of the South Coast Ranges. A series of low spots in the crest, more properly termed "water gaps" as remnants of old, uplifted river channels, occur here. From a climatological perspective, the heavier and cooler marine air masses flow through these gaps, across the range crest, and bring cooler marine air and sea breezes into the Paso Robles area. These gaps occur along a section of the Santa Lucia Range crest, bounded to the north by Rocky Butte and associated volcanic peaks, which reach elevations above 3,200 feet above sea level. To the south, the gaps are bounded by the Cuesta Pass and Lopez Mountain, reaching elevations over 2,800 feet above sea level. The range crest along the gap line is largely at 1,400 to 2,000 feet, allowing the incursion of marine air off the Pacific Ocean when the depth of the marine layer reaches those elevations, with spillover into the Paso Robles AVA. The lowest of these gaps is immediately west of the town of Templeton (see illustration below drawn by Nathanael Sheean under instruction by T.J. Rice; and Photo 14).

When the topography, elevations, land-sea breeze, and mountain-valley winds are combined as influences on local climates of the Paso Robles area, they set up gradients from west to east of adiabatic cooling (upslope on the west side of the Santa Lucia Range) and warming (downslope on the east side of the Santa Lucia Range), eastward marine incursion, a general westerly flow, bringing cooler air and sea breezes inland to the east. There is also a south to north gradient of local winds in the night versus daytime, as mountain winds move downslope, resulting in cold air drainage at night, with valley winds moving upslope and warming in the daytime. This is very important in much of the Paso Robles AVA, leading to lower early evening temperatures across the region and lower growing degree-day totals. There is also a less significant north to south incursion of marine air up the Salinas River Valley from Monterey Bay, reaching the areas of the proposed San Miguel District and Paso Robles Estrella District viticultural areas in the afternoon on the warmest of days in the summer and fall.

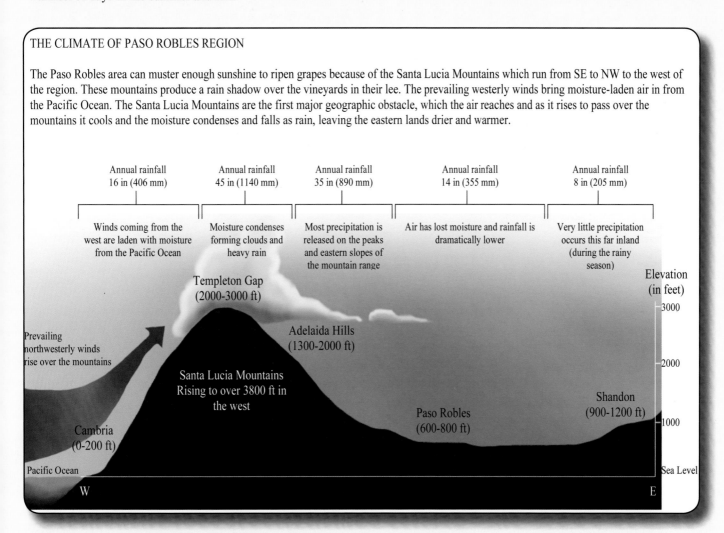

THE CLIMATE OF PASO ROBLES REGION

The Paso Robles area can muster enough sunshine to ripen grapes because of the Santa Lucia Mountains which run from SE to NW to the west of the region. These mountains produce a rain shadow over the vineyards in their lee. The prevailing westerly winds bring moisture-laden air in from the Pacific Ocean. The Santa Lucia Mountains are the first major geographic obstacle, which the air reaches and as it rises to pass over the mountains it cools and the moisture condenses and falls as rain, leaving the eastern lands drier and warmer.

The influence of mesoscale orography (mountains ranges and their topographic influences) has been modeled for the Central California Coast mountains. The model indicates airflow blocking below elevations of 500 meters (approximately 1,640 feet) with flow inland of cooler coastal air above this elevation.

Photo 14: Aerial view of the "Templeton Gap" in the Santa Lucia Range. Highway 46 West (bisecting photo) runs through the gap. Santa Rosa Creek Road is seen running diagonally at the top right. (24 March 2003)

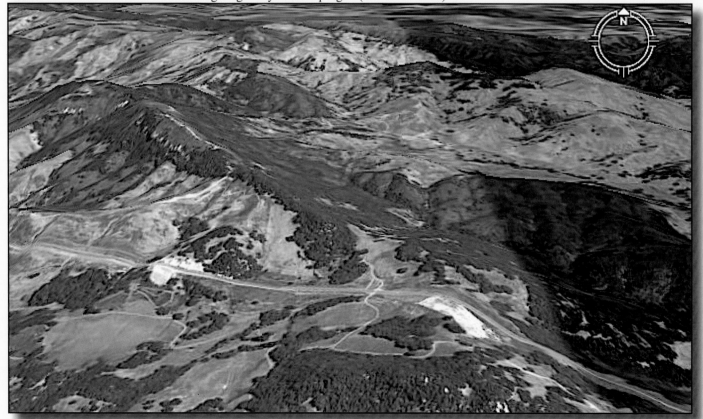

Google Earth image: Oblique aerial view of the "Templeton Gap" with Highway 46 West diagonally bisecting the image. Note that this image is a mosaic of two photos showing the western hills as greener (taken in the rainy season) as compared to the browner (taken in the dry season) easternmost slopes.

Paso Robles Airport Weather Station Climatic Data Summary

Table 5. Average temperature (°F) and precipitation (inches) data for the Paso Robles Airport (Years of observation: 1949 to 2005).

	Jan	Feb	Mar	Apr	May	Jun	Jul	Aug	Sep	Oct	Nov	Dec	Annual
Average Max. Temperature (°F)	61.10	64.00	67.80	73.70	80.00	86.60	92.70	92.90	89.30	81.10	69.30	61.70	76.70
Average Min. Temperature (°F)	33.00	36.40	38.50	40.20	44.10	47.70	50.80	50.00	47.40	41.90	35.70	32.30	41.50
Average Annual Temperature (°F)	47.05	50.20	53.15	56.95	62.05	67.15	71.75	71.45	68.35	61.50	52.50	47.00	59.10
Average Total Precipitation (in.)	3.15	3.16	2.40	1.05	0.28	0.05	0.03	0.04	0.19	0.55	1.39	2.63	14.91

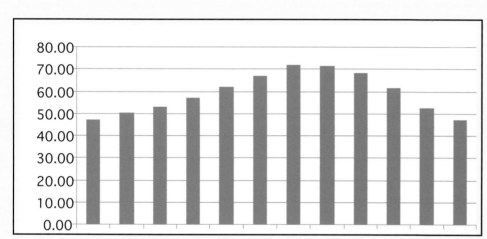

Figure 6. Average monthly air temperature (°F) at the Paso Robles airport weather station.

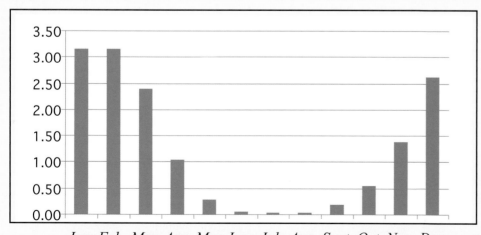

Figure 7. Average monthly precipitation (inches) at the Paso Robles airport weather station.

Soil climate and viticultural considerations

Climatic factors influence the supply of heat and light energy to the soil and vines. The "macroclimate" is the climate of the larger region (e.g., Mediterranean climatic zone). The "mesoclimate" is the climate of an individual vineyard site (e.g., an individual vineyard within the Paso Robles AVA). The "microclimate" is the climate within the grapevine itself.

Soil properties influence primarily the mesoclimate and microclimate of a vineyard (see photo to the right of Richard Smart and Gary Conway inspecting vines and weather station at Carmody McKnight vineyards; 8 June 2003). Darker colored soils will absorb more solar radiation during the day and the re-radiation of heat in the evening keeps the vines warmer for a longer period after sunset. Frost potential is reduced in these dark colored soil areas. Likewise, rock fragments at the soil surface will absorb solar energy and re-transmit the heat energy to the vines at night.

Climate data summary

The following table summarizes contemporary and historical climate data for the eleven proposed viticultural areas. From a geographic and viticultural perspective, the basic changes or gradients in climate across the Paso Robles AVA are set forth in this table shown below (adapted from Elliott-Fisk, 2007).

North to South	The north to south gradient shows temperature, but more importantly precipitation, differences, with the cooler and wetter area of the proposed Santa Margarita Ranch viticultural area to the south (receiving about 29 inches of precipitation annually) and warmer and the drier area of the proposed San Miguel District viticultural area to the north (receiving about 11 inches of precipitation a year).
West to East	The town of Shandon and the areas of Cholame and La Panza to the east are much drier and warmer than the areas of the proposed El Pomar District, Templeton Gap and Adelaida District viticultural areas and other locations to the westernmost parts of the AVA.
Mountain or Orographic Influence	The Tablas Creek weather station in the proposed Adelaida District records wetter conditions, along with the Templeton and Templeton Gap weather stations, and the city of Atascadero records intermediate precipitation values; the city of Paso Robles and the area of the proposed Paso Robles Estrella District viticultural area are slightly drier, though still intermediate in the range of annual precipitation values at the weather stations. The Shandon, San Juan Creek and Paso Robles Canyon Ranch areas are the driest zones.
Growing Degree Days	Growing degree days increase generally to the northeast (moving from the southwest) from the: (1) coolest Templeton Gap station to the higher locations of (2) Tablas Creek (Adelaida District) to (3) El Pomar District and Creston District on the high terrace bench east of the city of Atascadero and the town of Templeton to (4) Paso Robles Estrella District to San Miguel District, to (5) the high hills of the Geneseo District to (6) the town of Shandon and the area of San Juan Creek, and to (7) the Paso Robles Canyon Ranch areas of French Camp and Camatta Canyon in the southeast.

Relief (Topography)

The relief of the Paso Robles Area was determined mainly by past geological history interacting with the influences of climate. The area has three prominent physiographic land units: (1) the alluvial terraces, plains, and fans of the Salinas River and its tributaries; (2) the hills and mountains of the Santa Lucia and La Panza Ranges; and (3) the foothills of the Temblor Range and Cholame Hills.

All water drainageways eventually lead to the Salinas River. The Nacimiento River, Huerhuero Creek, and Estrella Creek are the main tributaries of the Salinas River. These water courses mainly dissect the Paso Robles Formation and adjacent geologic formations. The stream plains usually have by three to four levels of parallel terraces. Most of the weakly consolidated Paso Robles Formation has developed rounded geomorphic shapes.

In contrast to this gentle, rolling topography are the Santa Lucia and La Panza Ranges in the western and southern parts of the area. The greater weathering resistance of most of the rock parent materials and the relatively rapid regional tectonic uplift and faulting have caused long, relatively steep ridges to form. The ridgetops are often narrow, somewhat angular, and most have slopes of over 25 percent. Because of the steep and very steep slopes and the rock parent materials, most of the soils are well drained to somewhat excessively drained. Slope aspect differences (north vs. south) results in pronounced soil differences within these steep mountains and hills than within the Paso Robles area. Because north slopes generally have more soil moisture than south slopes, there is a greater buildup of organic matter and more intense mineral weathering. The Gaviota and San Andreas soils are examples of the effects of aspect on soil formation. The Gaviota soil, found on south slopes, is shallow and low in organic matter. The San Andreas soil is moderately deep and relatively high in organic matter on north aspects.

The topographic relief of a site will have a bearing on the surface and underlying soil hydrology (water movement). With undulating topography, suitable vineyards sites will typically be situated on the convex land surfaces. These features that tend to shed surface water, rather than collect it. Concave landforms, such as swales or gullies, are usually areas of water concentration and may have poorly drained soils. Depositional soils in these concave zones may be deeper due to erosion from higher areas. Thus, locating vineyards on convex landforms is one means of avoiding waterlogging of the vines.

Understanding the topography of a site involves analyzing soil parent material origins and determining the localized movement of water and soil materials. Topographic change indicates soil change and aids in determining where representative soils' backhoe pits should be excavated for soil morphologic descriptions and soil sampling for laboratory analyses.

The slope aspect is the compass direction that a slope faces. South and west facing slopes are the warmest vineyard sites in the earth's northern Hemisphere and are mainly grass vegetation in the Paso Robles AVA (see photo to the right with south-facing grasses); while north and east facing slopes are coolest and have mostly oak woodland vegetation (see photo to the right with north-facing oak woodland). Vineyards are often most desirable on south aspects, which face the sun during the growing season. In contrast, some white grape varieties have been successfully planted on north aspect slopes within the Paso Robles AVA. Slope steepness influences soil erosion rates. The steeper the slope, the greater the soil erosion potential. Water erosion can be reduced by using suitable cover crops, increasing soil organic matter levels, and increasing the surface soil roughness. Vineyards have been placed on slopes up to about 50% (fifty percent). Vineyard equipment use is limited on these steeper slopes and some slope terracing may be required to make them plantable to vines.

BIOLOGICAL ACTIVITY AND SOIL BIOLOGY

Vegetation, burrowing animals, insects, bacteria, and fungi are important contributors to the formation of soils. They cause gains or losses in organic matter, recycle plant essential nutrients, and cause changes in soil structure and porosity. Accumulation of organic matter has been an important process causing soil horizon differentiation in the Paso Robles area. The north aspects of the hills and mountains receive less direct sunlight. They support a relatively dense canopy of various broadleaf plants and hardwood trees, which result in about five percent humus and organic matter in these soils. This humus influences the dark color, structure, and physical condition of the soils. The Shimmon, Lompico, San Andreas, Andregg, and Linne soils are usually located on these north slopes. In contrast, the vegetation on the south slopes is mainly grasses and chaparral brush. This vegetative cover provides little shade and the soils are drier for longer periods, producing a less desirable habitat for microbial and vegetative growth. Organic matter is often less than one percent on south aspect soils. The Calleguas, Cieneba, Millsholm and Gaviota soils are on these south slopes.

As the rainfall decreases from the western mountains to the Adelaida Hills to the large eastern alluvial terraces, the soil organic matter levels decrease in most soils. This organic matter decrease affects the structure and porosity of the surface topsoil horizons.

The biological activity in the alluvial soils is related to soil texture and organic matter accumulation. The sandy Metz and Tujunga soils with lower water holding capacities do not naturally support lush plant communities and soil organic matter levels are less than one percent. In contrast, clay soils like Clear Lake in basins and are poorly drained with high water holding capacities. Consequently, they have accumulated higher amounts of organic matter. Water is available longer throughout the growing season on Clear Lake soil, and plants grow abundantly.

The native or natural vegetation results from the climate gradients across the Paso Robles AVA region and is depicted in the potential natural vegetation of California map (Kuchler, 1977). Mixed evergreen forest occurs in the mountain in the coolest and wettest climates, with blue-oak-grey pine woodland on slightly warmer and drier sites. Valley oak savannah occurs on the valley floors with higher water supplying capacities, California prairie is on drier valley floor sites, and alkali scrub is in the eastern, most arid sites. As elevations increase in the southeast, a conifer tree line is reached, with pines extending downslope into the alkali scrub and grassland.

"Healthy" soils contain a rich diversity of plant and animal species, most of which are microscopic (i.e., microbes) but some of which (e.g., earthworms) are easily observed. While some soil animals (like nematodes and phylloxera) and fungi cause diseases in grapevines, the vast majority of soil fauna and flora is essential to nutrient recycling and mineralization of organic matter (Baumgartner, 2003). The maintenance of this living aspect of the soil is essential to the maintenance of a healthy vineyard. Many of our farming practices (e.g., tillage, use of pesticides, crop monoculture, and soil compaction by machinery) tend to reduce biological diversity.

Laboratories can evaluate soil microbial diversity and provide an action plan to increase diversity if the soil bioassay is low for a particular soil organism functional group (*e.g.*, fungi and bacteria). Unfortunately, the interpretation of soil biological properties is an emerging science and is not sufficiently advanced to make cultural recommendations for potential and existing vineyards.

Nematodes and phylloxera should be characterized in the soil prior to vineyard establishment (see grape phylloxera, the yellow oval-shaped organisms, photo at the upper right). Sites that are in forests or that were previously planted to other fruit crops should be evaluated for the presence of nematodes and other soil-borne grape pathogens. Nematodes are small, wormlike parasites and several genera, notably *Xiphinema*, can transmit destructive viruses to grapevines (see a nematode photo at the lower right). Soil sampling for nematodes and the sample submission instructions can be obtained at local Agricultural Cooperative Extension offices. A passive pest control option involves planting and maintaining non-host plants, such as perennial grasses, on the site for up to several years before grapevines are planted to reduce the nematode populations. A more active approach involves planting a series of green manure crops, including a brassica (e.g., rapeseed, *Brassica napus*, "Dwarf Essex") that releases a chemical that is toxic to nematodes when the crop is incorporated into the soil. Detailed instructions on the use of this biocontrol measure can be obtained from Agricultural Cooperative Extension offices. Soil fumigation or treatment with nematicides is a third alternative, but one that carries potential risks, both to the user and to overall soil biology. Fumigation is often not recommended by agricultural advisors.

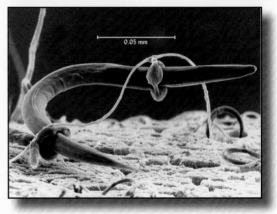

Other potential soil-borne pathogens include the oak root fungus (*Armillaria mellea*), which may infect grapevines in areas where affected oaks, or other hardwoods previously grew. Documented cases of oak root fungus disease have been recognized in California vineyards where oak trees are abundant or where oak tree roots persist after trees are removed (Merlander and Crawford, 1998).

TIME AND SOIL AGE

"Time zero," a soil's birthday, is considered to be that point in time when the present soil surface is exposed to the earth's atmospheric effects. The degree of mineral alteration (weathering) of parent material by the interacting forces of climate, living organisms, and relief is determined by the length of time these factors have acted on the soils. As a result of weathering intensity, older soils are typically located on stable upland geomorphic surfaces; whereas, the youngest soils are located on landscape positions that are subject to active erosion (such as landslides) or annual deposition (like floodplain channels). The oldest soils usually are those in which the parent material has been most altered by weathering.

Generally, the oldest soils are characterized by distinct boundaries between their horizons and have more recognizable soil horizons. Soils having few or no horizon differences are considered to be young. Soils having some or relatively indistinct horizon differences are considered to be of intermediate age. In contrast to the norm, some soils with few soil horizon differences are not necessarily the youngest. In these soils, the dominant influence of some other soil forming factor, such as highly resistant parent material, such as quartzite rock, may restrict soil horizon development. Entisols are the youngest soils, while Alfisols are considered the oldest soils in the Paso Robles AVA.

In general, the lowest alluvial stream channels and terraces consist of most recent alluvium, and the higher elevation terraces or fans are composed of the oldest alluvium. The age of the highest terrace or fan can be used to establish the range in age for various alluvial soils from the highest stream terrace to the lowest floodplain channel, such as in the Estrella River system.

The block diagram of the Nacimiento River area (see Figure 3) shows the topographic relationship of terrace soils in this area. In most places, at least three distinct terrace levels can be recognized, and time has influenced the amount of clay movement in the B horizon. The oldest soils have well developed subsoil B horizons enriched in clay, while the youngest soils lack subsoil B horizons.

California's Central Coast is geologically different from other California wine growing regions. Indeed, this region is geologically distinct from the rest of North America. This geological underpinning is a subject of great interest for us, because of its influence on the types of soils and mesoclimates found in the region, which accounts for distinctive vineyard growing conditions.

The San Andreas Fault, which cuts a jagged northwest-southeast fracture zone through California, forms the eastern border of the Central Coast. It also indicates the intersection of two vast tectonic plates - to the east, the North American Plate and to the west, the Pacific Plate. Thus, the rest of California, and virtually all of North America, sits on the North American Plate, the Central Coast is situated on the Pacific Plate.

The topography of California began to be defined 130 million years ago (Ma) when a tectonic plate called the Farallon Plate slid under the North American continent into a deep ocean trench (see Figure 8). The significance for viticulture is that, as the plate descended, thick layers of ocean sediment were scrapped on to the continental plate. This ancient process set into motion a chain of events that was to result in the diversity of geology and resultant soils found in the Central Coast today.

The principal geology of the Central Coast was subsequently created about 30 million years ago when the North American and Pacific plates met and began to slip laterally past each other. As the San Andreas Fault developed along their boundary, a large shelf of crustal rock was torn off the Pacific Plate, creating a mass of basement rock called the Salinian Block, which today underlies much of the Central Coast. The Salinian Block consists of hard granitic rocks, such as quartz diorite, forged deep beneath the earth's surface and metamorphic rocks, such as serpentinite, formed by the partial melting of the sea floor crust.

At the southern geologic border of the Central Coast are the Transverse Ranges. They were formed from compression and uplift as the Farallon Plate slid under the North American Plate. This mountain range runs east-west; in contrast to the prevailing north-south trend of most of the mountain ranges in North America. The main effect of the coast-parallel faulting patterns along the California coast was to establish a series of submarine basins in areas that were formerly above sea level. The first sediments deposited in the basins were coarse and thin, developed along shorelines in shallow ocean waters. These marine sediments hardened into sedimentary rocks, which were previously named the Tierra Redondo Formation, but are now called the Vaqueros Formation.

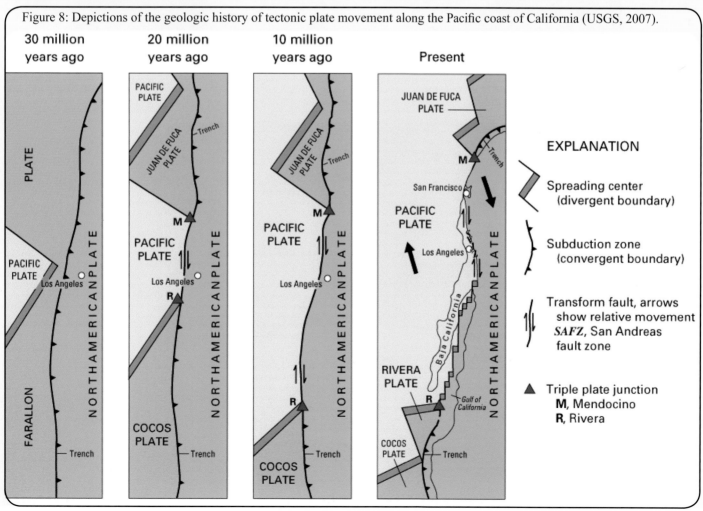

Figure 8: Depictions of the geologic history of tectonic plate movement along the Pacific coast of California (USGS, 2007).

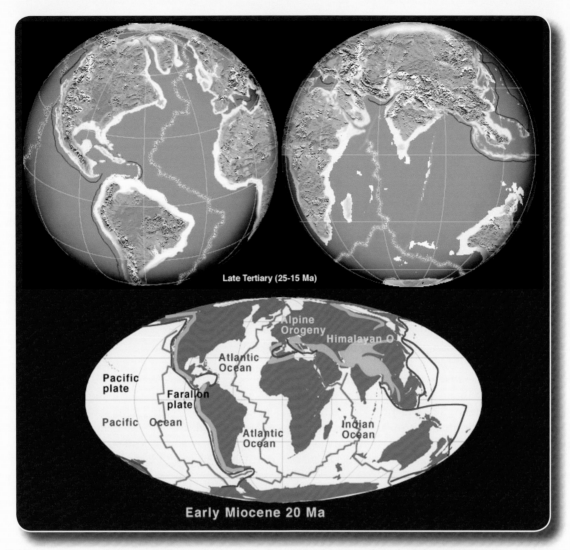

Figures 9-10: Depictions of the earth during the Late Tertiary (25-15 Ma) and Early Miocene (20 Ma) as compared to the "Modern World." The Paso Robles AVA was submerged under the Pacific Ocean during the Miocene Epoch (Blakey, 2007).

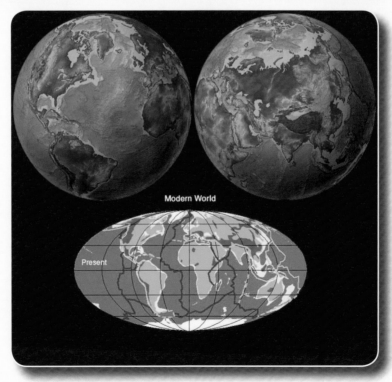

As the Oligocene Epoch ended, and the Miocene began, there was an episode of intense volcanic activity. Over 200 feet of volcanic ash deposited and hardened into tuff, which is named the Obispo Formation. The Obispo Formation was followed by the deposition of sediments in ocean depths of 2,000 to 7,000 feet.

These thinly bedded calcareous and siliceous sediments formed into shale, mudstone and siltstone rocks, which are now called the Monterey Formation. The sediment sources were muds from the land, combined with siliceous diatoms and other marine plankton. The Coast Ranges were largely submerged below the ocean at this time. Environmental conditions resembled those that now exist along the Southern California Borderland, a faulted shelf of deep basins and uplifted islands that lies between Santa Barbara and San Diego. Sea floor depths were variable among the submarine basins. The resulting sediment chemistry and grain size also varied relative to the proximity to the land and the effects of ocean current flow. Calcareous and coarser sediments were likely deposited in shallower ocean waters closer to the land-sea margins.

In the late Miocene, as the ocean level was becoming more shallow, the Santa Margarita Formation sediments were being deposited. These shallow marine deposits of the Santa Margarita Formation contain abundant surf-zone fossil beds that resemble "fossil reefs" of clam and oyster shells, which were piled up by storms along beach margins (Chipping, 1987).

The late Pliocene and Pleistocene Epochs were marked by a time of mountain building that continues today. The Miocene and early Pliocene rocks were folded and faulted and, in some cases, pushed upward several thousand feet. At the same time, the ocean levels fell, as glaciers formed in the Sierra Nevada during the Pleistocene Ice Ages.

The land uplift was slow enough that broad erosional surfaces were established on the tops of the folded rocks, sloping eastward away from a central uplifted ridge spine along what is now called the Cuesta Ridge. The remains of these surfaces can be seen as the upper elevations of the Irish Hills southwest of San Luis Obispo.

Sediments from the eroding new mountains spread eastward and southward from the Santa Lucia Range as alluvial fans were constructed. In the Paso Robles AVA region, the mixed gravelly and finer-textured sediments in these fans are now called the Paso Robles Formation; the name now assigned to any alluvial sediment of about the same age. The deposited sediments now cover the land surfaces between Paso Robles and Shandon and bury a variety of older rock formations. The alluvial fans have been tilted, so that the water courses today flow in nearly opposite directions as compared with the Salinas River drainage system (Chipping, 1987; Dibblee, 2004).

GEOLOGY OF THE PASO ROBLES GROUNDWATER BASIN AND WITHIN THE PASO ROBLES AVA

WATER-BEARING GEOLOGIC FORMATIONS (text modified from Fugro et al., 2002)

ALLUVIUM

Alluvial deposits occur beneath the floodplains of the rivers and streams within the basin. These deposits reach a depth of about 100 feet below ground surface (bgs) or less and are typically comprised of coarse sand and gravel. The alluvium is generally coarser than the Paso Robles Formation sediments, with higher permeability that results in well production capability that often exceeds 1,000 gallons per minute (gpm). The principal areas of groundwater recharge to the basin occur where the shallow alluvial sand and gravel beds are in direct contact with the Paso Robles Formation.

PASO ROBLES FORMATION

The Paso Robles Formation (Qtp) is a Plio-Pleistocene, predominantly nonmarine geologic unit comprised of relatively thin, often discontinuous sand and gravel layers interbedded with thicker layers of silt and clay. It was deposited in alluvial fan, flood plain, and lake depositional environments. The formation is typically unconsolidated and generally poorly sorted. It is not usually intensely deformed, except locally near fault zones. The sand and gravel beds within the unit have a high percentage of Monterey Formation shale gravel and generally have moderately lower permeability compared to the shallow, unconsolidated alluvial sand and gravel beds. The formation is typically sufficiently thick such that water wells generally produce several hundred gpm. In the area near Atascadero, the Paso Robles Formation has been folded, exposing the basal gravel beds. With the basal gravel exposed and in direct contact with the shallow alluvium, the Paso Robles Formation is recharged directly from the river alluvium (see Figure 11).

The Paso Robles Groundwater Basin is comprised predominantly of Paso Robles Formation sedimentary layers that extend from the ground surface to more than 2,000 feet below sea level in some areas (resulting in basin sediments with a thickness of more than 2,500 feet). Throughout most of the basin, however, the water-bearing sediments have a thickness of 700 to 1,200 feet (with the base of the sediments more or less at sea level). See the geologic map of the Paso Robles groundwater basin on the next page (this map is adapted from Fugro et al., 2005).

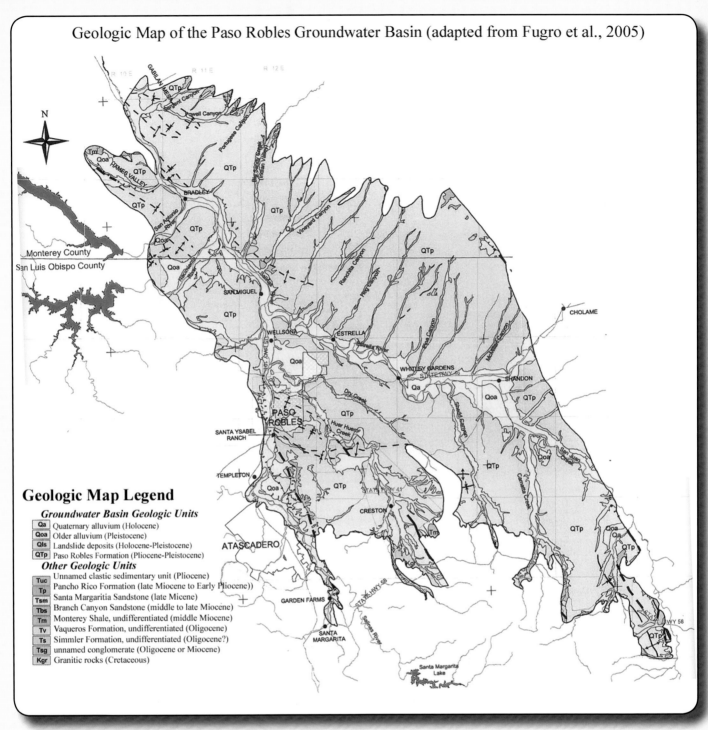

Geologic Map of the Paso Robles Groundwater Basin (adapted from Fugro et al., 2005)

Geologic Map Legend

Groundwater Basin Geologic Units

Qa	Quaternary alluvium (Holocene)
Qoa	Older alluvium (Pleistocene)
Qls	Landslide deposits (Holocene-Pleistocene)
QTp	Paso Robles Formation (Pliocene-Pleistocene)

Other Geologic Units

Tuc	Unnamed clastic sedimentary unit (Pliocene)
Tp	Pancho Rico Formation (late Miocene to Early Pliocene))
Tsm	Santa Margaritia Sandstone (late Micene)
Tbs	Branch Canyon Sandstone (middle to late Miocene)
Tm	Monterey Shale, undifferentiated (middle Miocene)
Tv	Vaqueros Formation, undifferentiated (Oligocene)
Ts	Simmler Formation, undifferentiated (Oligocene?)
Tsg	unnamed conglomerate (Oligocene or Miocene)
Kgr	Granitic rocks (Cretaceous)

GEOLOGIC FORMATIONS WITH LOWER WATER-BEARING CAPACITIES

Underlying the basin sedimentary beds are older geologic formations that typically have lower permeability and/or porosity. In some cases, these older beds yield in excess of 50 gpm flow to wells but they often have poor quality water or are of limited extent, such as are found along a fault fracture zone. In general, the geologic units underlying the basin include Tertiary-age consolidated sedimentary beds, Cretaceous-age metamorphic rocks, and granitic rock.

TERTIARY-AGE CONSOLIDATED SEDIMENTARY FORMATIONS

The Tertiary-age older consolidated sedimentary formations include the Pancho Rico Formation, an unnamed clastic unit, the Santa Margarita Formation, the Monterey Formation, the Obispo Formation, and the Vaqueros Formation. These units crop out around most of the basin edge and underlie the basin sediments.

The Pancho Rico Formation (Tp) is a Pliocene-age marine deposit found mostly in the northern portion of the study area. In places, it appears to be time-correlative to the Paso Robles Formation, and may be in lateral contact as a facies change. The unit is predominantly

Figure 11: Geologic Map of Stephan Vineyard and L'Aventure Winery, which is located in the southwest corner of Section 12, Township 27 South, Range 11 East, Mount Diablo Baseline and Meridian, California. The Monterey Formation, Sandholdt member (Tml) is the major geologic formation in the western part of this area along with recent alluvial deposits (Qa), which cover the sedimentary rocks in the valley bottoms. The Paso Robles Formation (QTp), the older alluvium of Pleistocene age, is located in the eastern part of this map and merges with recent Salinas River alluvium (Qa), which is located further east of this map (Dibblee, 2004e).

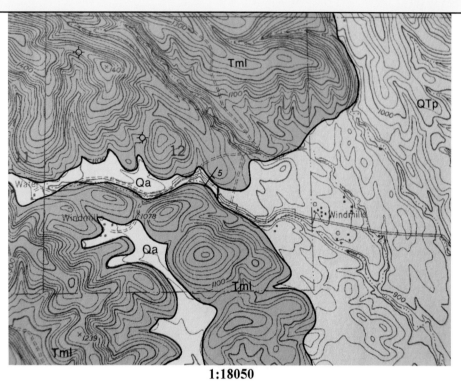

1:18050
20 ft contour interval

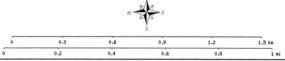

Geologic Map Unit Key:

Qa: Alluvial gravel, sand and clay

Tml: Lower part (Sandholdt member of Durham, 1968) shale, thin bedded, semisiliceous, platy to soft, fissile, punky, cream-white weathered, includes thin layers of yellow-gray dolomite, shale contains foraminifera and fish scales diagnostic of middle Miocene age (Luisian-Rilizian Stage, Smith and Durham, 1968).

Sandholdt member of Monterey Formation (Miocene) – Includes:
Calcareous and porcelaneous mudstone – Thin- to thick-bedded, chocolate-brown to buff, calcareous, foraminifera mudstone, most abundant in lower part of section; locally phosphatic and glauconitic. Thin- to medium-bedded, porcelaneous mudstone most abundant in upper part of exposed section; weathers to light-gray rock of low density locally known as "chalk-rock". Locally abundant concretions, lenses, and rare beds of buff to grayish-orange limestone and dolomite. Rare dark-gray cherty mudstone and thick lenses of laminated dark-gray chert. Local graded sandstone beds. About 400 m thickness exposed.

(Note: The Qa and Tml geologic map units are located within Stephan Vineyard. The QTp map unit is outside the boundaries of Stephan Vineyard.)

comprised of fine-grained sediments up to 1,400 feet thick that yield low quantities of water in the Gabilan Mesa area north of the Paso Robles basin. The upper Miocene-age unnamed clastic unit (Tuc) is time-equivalent to the Pancho Rico Formation and is comprised of up to 200 feet of sandy conglomerate beds in the Shandon area. This unit is cemented and produces limited flow to wells.

The Santa Margarita Formation (Tsm) is an upper Miocene-age marine deposit, consisting of a white, fine-grained sandstone and

siltstone with a thickness of up to 1,400 feet. The unit is found beneath most of the groundwater basin. The Santa Margarita Formation crops out in the Santa Margarita area where more than 300 domestic water wells depend on its very limited flow capabilities. It is also a host to a number of springs. South of Templeton, water produced from the Santa Margarita Formation is often of acceptable water quality. However, north of Templeton in the area beneath the City of Paso Robles, the unit becomes progressively more permeable and is the main reservoir for the historical presence of geothermal water. Groundwater in the geothermal areas is often under pressure and artesian flow is a common occurrence, with flow rates at times exceeding 400 gpm. The Santa Margarita Formation aquifer in this area is not considered part of the Paso Robles basin because the produced water quality is usually very poor. The geothermal waters contained in the Santa Margarita Formation in this area are often highly mineralized and characterized by elevated boron concentrations that restrict agricultural uses. North of the study area, the Santa Margarita Formation crops out in the upper portions of the Gabilan Mesa. South of the basin, it is exposed along Highway 58 where springs occasionally issue from the unit.

The Miocene-age Monterey Formation (Tms or Tml) consists of interbedded argillaceous, calcareous, and siliceous shale, sandstone, siltstone, and diatomite (see Figure 12). Some dolomitic and calcitic limestone beds from 0.5 to 2.0 feet thickness are found as summit cap rocks in the Adelaida Hills district (see the upper right photo of calcareous shale at Carmody McKnight vineyards). This rock unit is as great as 2,000 feet thick in the Paso Robles groundwater basin area, and is often highly deformed. It is exposed south and west of the groundwater basin. Water wells completed in the Monterey Formation may be quite productive if a sufficient thickness of highly deformed and brittle siliceous shale is encountered (see the lower right photo showing outcrops of siliceous shale from the Monterey Formation along Peachy Canyon Road). Springs issue from the Monterey Formation in the Atascadero area and on Cuesta Ridge south of the basin. The Monterey Formation is also a source for oil as well as water in the area near Hames Valley, downstream of Lake San Antonio, and in upper Indian Valley. Groundwater produced from the Monterey Formation often has high concentrations of hydrogen sulfide, total organic carbon, and manganese. In the Paso Robles area, the Monterey Formation may be a source of geothermal water that has high sulfide concentrations in addition to high boron, iron, manganese, and total dissolved solids.

The lower Miocene-age Obispo Formation (To) is not found adjacent to or underlying the Paso Robles Groundwater Basin sediments but is described here briefly because it is found in the watershed south of the study area. It is a consolidated volcanic tuff bed underlying the Monterey Formation south of Santa Margarita. Wells in the Tassajara Creek area produce more than 100 gpm from the formation, and it is the host of several springs along Cuesta Ridge. Water produced from this unit is generally moderately saline (total dissolved solids concentrations around 1,000 mg/l).

Figure 12: Geologic map of the Adelaida (Lincoln) School area (NW corner of map) and Halter Ranch (SW portion; Durham, 1983).

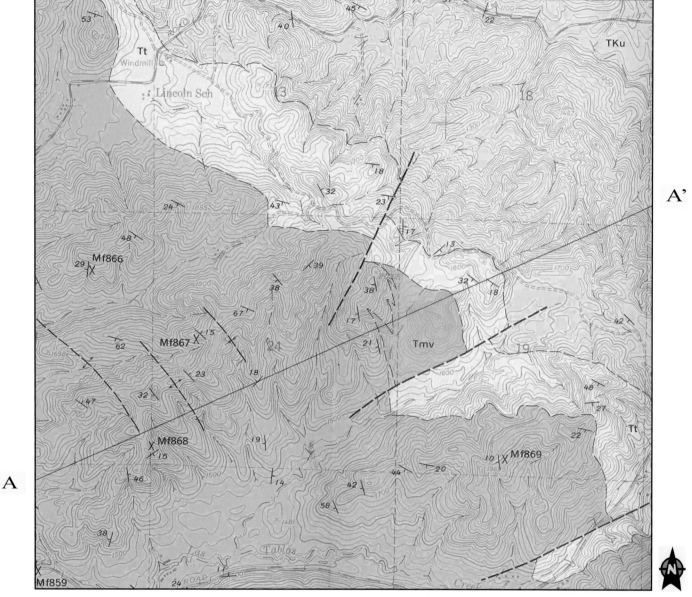

Map Legend:

Geologic cross section (A to A')

Scale: 1:20,000
Contour interval: 20 feet

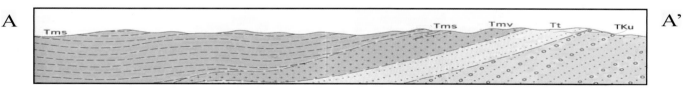

- **_Tms: Monterey Formation, Sandholdt member (Miocene age)._** Sandholdt member, chiefly mudstone and shale, well-indurated, calcareous. Very pale orange or yellowish grey, abundant Miocene plankton and fish scales, rare mammal fossils. Dolomitic limestone carbonate rock, beds 0.5 to 2 feet thick, yellowish-grey or grayish-orange. Siliceous and cherty rocks are associated.
- **_Tmv: Volcanic (igneous) rocks (Miocene age)._** Olivine basalt and gabbro, dark-greenish-grey, poorly preserved pillow structure, weathered, forms moderate orange-brown soil; contributes sandy alluvium and colluvium to nearby calcareous rocks of Monterey Formation, Sandholdt member. Igneous intrusion into the Monterey sedimentary rocks.
- **_Tt: Tierra Redondo (Vaqueros) Formation (Miocene age)._** Sandstone, fine- to medium-grained, some conglomerate, well sorted to poorly sorted, friable to well cemented, calcareous and noncalcareous, generally massive and white. Conglomerate, pebbles, cobbles of granitic, metamorphic, and volcanic rocks, well cemented, massive, white or yellowish-grey.
- **_TKu: Atascadero Formation (Cretaceous age)._** Sandstone, medium to coarse grained, arkosic, biotite common, calcareous to siliceous; Conglomerate, pebbles, cobbles and boulders derived from igneous and metamorphic rocks, noncalcareous; Mudstone, well indurated, noncalcareous, sedimentary beds 1 - 6 inches thick; and Siltstone, well indurated, noncalcareous, poorly bedded, plant remains are conspicuous on some bedding surfaces. (**NOTE**: " - - - - " represents a fault line).

The marine Oligocene-age Vaqueros Formation (Tv map unit) (also called Tierra Redondo Formation, Tt map unit, by Durham, 1968) is a highly cemented arkosic and fossiliferous sandstone that reaches a thickness up to 200 feet (Dibblee, 1971; see Figure 12 showing the close association of the three geologic rock types, shale, sandstone and basalt, on Halter Ranch East and photo to the right of Vaqueros sandstone on the Matilija Vineyard located on Adelaida Road across from Adelaida Cellars tasting room and walnut orchard). Some conglomerate beds are associated with the sandstone in some areas. Springs with flows up to 25 gpm are common in canyons on the western and southern sides of the basin. Most water wells tapping this formation produce less than 20 gpm. Generally, the quality of water in this unit is good, bit it is hard due to the calcareous nature of this rock.

The Paleocene to Upper Cretaceous-age Atascadero Formation is a thick-bedded, light gray to light brown, coherent sandstone. It is arkosic and micaceous with lenses of cobble conglomerate of andesitic, volcanic and plutonic rocks and thin lenses of micaceous claystone (Dibblee, 1971).

The southern and western edges of the groundwater basin are bordered by Cretaceous-age metamorphic and granitic rock. The metamorphic and sedimentary rock units include the Franciscan, Toro, and Atascadero Formations. The Franciscan Formation consists of discontinuous outcrops of shale, chert, metavolcanics, graywacke, and blue schist, with or without serpentinite. The Franciscan Formation has an undetermined thickness and has low permeability and porosity. Limited volumes of groundwater can be produced from this geologic unit, generally only where the metavolcanics rock has been highly fractured.

Photos : Olivine basalt and gabbro (*Tmv* map unit symbol on geology map) weathered on a hillslope and summit at Carmody McKnight vineyards and soil profile (Linne variant clay loam) weathered from these mafic (high Fe and Mg) igneous rocks.

The Toro Formation (Kt) is a highly consolidated claystone and shale that does not typically yield significant water to wells. The Atascadero Formation (Ka) is highly consolidated but does have some sandstone beds that yield limited amounts of water to wells. Both the Toro and Atascadero formations are exposed in the hills west of Santa Margarita, Atascadero, and Templeton.

The granitic rock (Kgr) lies east of the Rinconada Fault zone, south of Creston, east of Atascadero, and in the area northwest of Paso Robles. The Park Hill area south of Creston and east of Atascadero is well known for the difficulty of finding sufficient groundwater to serve single family residences. Where water is found, it is typically low in salinity. The granitic rocks often have a decomposed regolith up to 80 feet in thickness in the valley floor areas that may contain limited amounts of groundwater despite low sediment permeability due to the breakdown of feldspar and iron and magnesium silicates into clays and fine grained sediment.

Table 6: Major geologic units and their associated rocks and sediments within the Paso Robles AVA.

GEOLOGIC UNITS	PREDOMINANT ASSOCIATED ROCKS & SEDIMENTS
Atascadero Formation	Sandstone, shale (metamorphosed)
Cuesta Ophiolite	Serpentinite, peridotite, gabbro, basalt, chert, shale, sandstone
Granite	Granite with quartz intrusions; diabase
Franciscan Formation	Peridotite, gabbro, basalt, diabase, chert, shale, sandstone, graywacke, blue schist, conglomerate, serpentinite
Monterey Formation	Shale, mudstone, siltstone, dolomite, calcitic limestone
Obispo Formation	Volcanic tuff
Pancho Rico Formation	Stratified fine- to coarse-grained marine sediments
Paso Robles Formation	Stratified alluvial sand, gravel and clays
Rincon Formation	Sandstone, siltstone, shale, conglomerate, dolomite nodules
Santa Margarita Formation	Sandstone, siltstone
Serpentinite	Serpentinite, peridotite
Tierra Redondo Formation	Sandstone, conglomerate
Toro Formation	Shale, sandstone
Vaqueros Formation	Sandstone, conglomerate
Volcanics	Volcanic diabase, basalt, gabbro

Table 7: Geologic age timetable for rocks and sediments within the Paso Robles AVA.

ERA	PERIOD	EPOCH	AGE (years) (Millions B.P.)	ROCK FORMATIONS OR UNITS
Cenozoic	Quaternary	Holocene	0.01 to present	Recent alluvial deposits
		Pleistocene	0.01 to 1.8	Paso Robles
	Tertiary	Pliocene	1.8 to 5.3	Pancho Rico, Monterey (Sandholdt)
		Miocene	5.3 to 23.8	Monterey, Obispo, Rincon, Santa Margarita, Vaqueros (Tierra Redondo)
		Oligocene	23.8 to 33.7	Vaqueros
		Eocene	33.7 to 54.8	none
		Paleocene	54.8 to 65	Atascadero
Mesozoic	Cretaceous		65 to 144	Cuesta Ophiolite, Franciscan, Toro, serpentinite, and volcanics
	Jurassic		144 to 206	Toro and possibly serpentinite and volcanics

Table 8: Abundant rocks within the Paso Robles AVA (organized by rock type).

IGNEOUS	*SEDIMENTARY*	*METAMORPHIC*
Basalt	Chert	Blue schist
Diabase	Conglomerate	Gneiss
Gabbro	Graywacke	Meta-sedimentary
Diorite; Granite	Limestone (photo to right); Dolomite	Meta-igneous
Peridotite	Mudstone	Serpentinite
Tuff	Siltstone/Shale	Slate

Photo 15: Tablas Creek vineyards on soils derived from Monterey Formation rocks (viewed from Adelaida Road; 24 July 2007).

IRRIGATION RELATED WATER CONDITIONS WITHIN THE PASO ROBLES AVA

FINAL REPORT PASO ROBLES GROUNDWATER BASIN (Fugro et al., 2002)
(Internet Source: http://www.slocountywater.org/site/Water Resources/Reports/Paso Phase 1/index.htm).

The executive summary from this Paso Robles groundwater basin study is found below with few minor corrections to the original report text. This summary is already a concise explanation of the groundwater basin conditions and we decided not to further summarize and edit the groundwater basin report. The largely unedited text is shown below within quotation marks. Readers who wish to reference the complete groundwater basin reports should refer to the web site listed above and its follow-up Phase II study (Internet Source: http://www.slocountywater.org/site/Water Resources/Reports/Paso Phase 2/index.htm).

"EXECUTIVE SUMMARY

GENERAL

This Final Report of the Paso Robles Groundwater Basin study presents the results of efforts to investigate and quantify the hydrogeologic conditions of the basin. The work was conducted jointly by Fugro West, Inc. and Cleath and Associates, in conjunction with Peter Canessa, P.E. and ETIC Engineering, Inc.

The Paso Robles Groundwater Basin study was a technical investigation intended to provide the San Luis Obispo County Public Works Department, North County public water agencies, and overlying landowners and water users a better understanding of the basin by answering questions related to the quantity of groundwater in the basin, the hydraulic movement of groundwater through the basin, sources and volumes of natural recharge, and trends in water quality.

BASIN DEFINITIONS AND BASIN BOUNDARIES

The Paso Robles Groundwater Basin encompasses an area of approximately 505,000 acres (790 square miles) in both San Luis Obispo and Monterey counties. The basin ranges from the Garden Farms area south of Atascadero to San Ardo in Monterey County, and from the Highway 101 corridor east to Shandon. Internally, a single hydrologically distinct subbasin was defined. The Atascadero subbasin encompasses the Salinas River corridor area south of Paso Robles, including the communities of Garden Farms, Atascadero, and Templeton.

GROUNDWATER OCCURRENCE, LEVELS AND CONDITIONS

Water level data showed that from July 1980 through June 1997 there is no definitive upward or downward water level trend for the whole basin. However, different water level trends are observed at specific locations in the basin. Water levels have declined, in some areas rather dramatically, in the Estrella and San Juan Creek areas, with rising water levels in the Creston area.

In general, groundwater flow moves northwesterly across the basin towards the Estrella area, thence northerly towards the basin outlet at San Ardo. The biggest change in groundwater flow patterns during the base period is the hydraulic gradient east of Paso Robles, along the Highway 46 corridor, which has steepened in response to greater pumping by the increasingly concentrated development of rural ranchettes, vineyards, and golf courses.

WATER QUALITY

In general, the quality of groundwater in the basin is relatively good, with few areas of poor quality and few significant trends of ongoing deterioration of water quality. Historical water quality trends were evaluated to identify areas of deteriorating water quality. A major water quality trend is defined as a clear trend that would result in a change in the potential use of water within 50 years, if continued.

Six (6) major trends of water quality deterioration in the basin were identified, including:

1.) increasing total dissolved solids (TDS) and chlorides in shallow Paso Robles Formation deposits along the Salinas River in the central Atascadero subbasin;
2.) increasing chlorides in the deep, historically artesian aquifer northeast of Creston;
3.) increasing TDS and chlorides near San Miguel;
4.) increasing nitrates in the Paso Robles Formation in the area north of Highway 46, between Salinas River and Huerhuero Creek;
5.) increasing nitrates in the Paso Robles Formation in the area south of San Miguel; and
6.) increasing TDS and chlorides in deeper aquifers near the confluence of the Salinas and Nacimiento rivers.

GROUNDWATER IN STORAGE

The total estimated groundwater in storage within the Paso Robles Groundwater Basin is approximately 30,500,000 acre feet (af). This value changes yearly, depending on recharge and net pumpage. Between 1980 and 1997, groundwater in storage increased approximately 12,400 af, an approximate 0.04% increase. This represents an average increase in storage of 700 acre feet per year (afy). On one hand, this relatively small percentage could be viewed as an indication of stable basin-wide conditions. However, it is noted that steadily decreasing storage in the 1980's was offset by increased water in storage throughout the 1990's. Furthermore, not all areas of the basin have observed the same trends in water levels and change in storage. In the Atascadero subbasin, total groundwater in storage averaged about 514,000 af. Approximately 2,600 af more groundwater was in storage in the subbasin in 1997 compared to 1980, a 0.5% increase in total groundwater in storage during the base period. This represents an increase of about 200 afy in storage.

HYDROLOGIC BUDGET

The purpose of a hydrologic budget (or water balance) is to assess all the inflows and outflows of water to the groundwater basin over the base period. The water budget was performed by calculating each component of water inflow and outflow for each year of the base period, and comparing the totals to the annual change in groundwater in storage as determined by the specific yield method. The base period, defined in this study from July 1980 through June 1997, is a representation of the long-term conditions of water supply.

The hydrologic budget is simply a statement of the balance of total water gains and losses from the basin, and is summarized using an equation referenced in the report (Fugro et al., 2002). Using this inventory, the sum of all the components of outflow from the Paso Robles Groundwater Basin exceeded the sum of all the components of inflow by an estimated 2,700 afy. As described earlier, an independent method of calculating the change in the volume of groundwater in storage was performed using the specific yield method and compared to the results of the inventory method. This approach indicated a slight annual increase in groundwater in storage of about 700 afy. For the Atascadero Subbasin, the sum of all the components of outflow approximately equaled inflow during the base period, with total groundwater in storage of about 514,000 af.

The change in storage calculation showed an annual increase in groundwater over the 17-year base period of about 200 afy. Reconciliation of the hydrologic budget shows a consistency in the results of the two methods of calculation. At first glance, the results of the hydrologic budget calculations, along with the change in storage calculations and analysis of the water level data, indicate a basin-wide stability. This conclusion, however, is tempered by the recognition that parts of the basin have experienced significant declines in water level over the past several years, particularly in the Estrella area along the Highway 46 corridor from the eastern edge of Paso Robles to Whitley Gardens as a result of relatively concentrated development of rural residential housing, golf courses, and vineyards.

PERENNIAL YIELD

The perennial yield of a basin, as defined in this investigation, is the rate at which water can be pumped over a long-term without decreasing the groundwater in storage. Many definitions of perennial yield (or safe yield) tie the concept of basin yield to the rate of groundwater extraction that will not create an economic impact. However, for the purposes of the groundwater basin study, the concept of perennial yield is more closely tied to the natural rate of replenishment or recharge to the basin, such that there is no decrease in groundwater in storage.

The results of this investigation indicate a perennial yield value of approximately 94,000 afy for the Paso Robles Groundwater Basin (which includes the Atascadero subbasin). By calculation, the perennial yield of the Atascadero subbasin approximates 16,500 afy.

BASIN CONDITIONS IN 2000

In the year 2000, groundwater pumpage in the Paso Robles Groundwater Basin was approximately 82,600 af, compared with the perennial yield estimate of 94,000 afy. Similarly, Atascadero subbasin pumpage in the year 2000 was approximately 11,100 af, compared to the perennial yield estimate of 16,500 afy.

Total net groundwater pumpage in the basin (and the subbasin) declined steadily from 1984 through 1998. Groundwater production data since 1998 show, however, that groundwater pumpage may again be increasing. Pumpage in 2000 was higher than at any previous time since 1992. It should also be noted that groundwater pumpage exceeded the perennial yield from the start of the base period in 1980 through 1990. Only in the last decade has pumpage been less than the perennial yield.

Currently, agricultural pumpage comprises 69% of total basin pumpage. Depending on new trends or pressures in the agricultural industry, it is likely that basin pumpage will approach or exceed the perennial yield in the near future. The San Luis Obispo County Master Water Plan Update projects future water demands for the area to be 120,620 afy by the year 2020, which suggests that future water demands may soon exceed the 94,000 afy perennial yield of the basin.

In the Atascadero subbasin, municipal, rural domestic, and small commercial water systems comprise 91% of total pumpage in the subbasin. Interpolation of data from the County Master Water Plan projects water demand in 2020 in the Atascadero subbasin to be in the range of 16,000 to 20,000 afy, compared to the perennial yield value of 16,500 afy.

It is important to note that short-term periods of groundwater extractions in excess of the perennial yield will not necessarily result in significant negative economic impacts. Groundwater in storage in the basin is sufficiently large such that short-term overdraft conditions may be acceptable to withstand drought periods.

RECOMMENDATIONS FROM THE GROUNDWATER BASIN STUDY

It is recommended that a basin-wide numerical groundwater flow model be developed for the Paso Robles Groundwater Basin. The model will serve as a tool for quantitative evaluation of existing and future hydraulic conditions across the basin, including changing groundwater level elevations, well yields, natural and artificial recharge, and associated effects on surface water-groundwater interaction and water quality.

Specifically, the objectives of the model include:

Refining uncertain components of the hydrologic budget for the basin;

Refining estimates of perennial yield for the basin;

Evaluating water quality trends in response to hydraulic changes across the basin;

Evaluating potential impacts on groundwater levels and perennial yield as a result of continued and varied basin operations and hydraulic conditions; and

Defining operational options for comprehensive and/or localized management of groundwater use across the basin (this is the end of information borrowed from Fugro et al., 2002)."

GROUNDWATER SOURCES OUTSIDE OF THE PASO ROBLES GROUNDWATER BASIN

Many productive vineyards in the Paso Robles AVA are located outside of the defined boundaries of the groundwater basin. Most of these vineyards are located west of the basin. In these vineyards, the viticulturists rely on irrigation water directly delivered from groundwater wells (see photo to the right of a water well on Carmody McKnight vineyards) and, in some cases, from surface reservoirs that are constructed to store groundwater along with natural precipitation.

The groundwater sources in these areas are aquifers that are mainly associated with fractured rock zones but not with deep alluvium, as is the case in the groundwater basin. The subsurface fractured rock zones are often not easily located by hydrogeologists. Most wells are drilled to depths of over 700 feet in order to find sufficient quantities of groundwater (personal communication; Steve Vierra, 2005; Mitch Wyss, 2007). Often, water is not immediately found and multiple drill holes are required to find acceptable groundwater sources (personal observations; former DeBro vineyard, 1997).

The fact that many vineyards have experienced reduced water quantities in their irrigation wells implies that the aquifers in these rock fracture zones do not receive sufficient annual recharge water to replace the annual irrigation water demands (personal observations; Adelaida Hills landowners, 2007). The groundwater can be considered to be "relict water," which may never be realistically replaced. Consequently, most vineyard land sales contracts in these areas have a stipulation that the irrigation water supply must be of sufficient quality and quality to sustain a productive vineyard over a reasonable period of time (20 years or more). Some land sales have collapsed due to the inability of the seller to meet this requirement. Long-term written records do not exist for most of the irrigation water wells both within and outside of the groundwater basin (see photo to the right of an agricultural water well at the intersection of Wellsona and Airport roads, north of Paso Robles). Therefore, it is very difficult to accurately predict the long-term groundwater supply situation, especially in the rock fracture zones outside of the groundwater basin. In the future, an adequate supply of good quality irrigation water may be the primary limiting factor controlling the expansion of productive vineyards located outside of the groundwater basin in the larger Paso Robles AVA.

ADDITIONAL INFORMATION SOURCES FOR CLIMATE AND WATER IN THE PASO ROBLES AVA

The San Luis Obispo County Water Resources, Division of Public Works, maintains a web site, which lists information and services related to hydraulic planning, major projects, water quality lab, water resources, and live weather data (Internet Source: http://www. slocountywater.com)

The Paso Robles Wine Country Alliance (WCA) offers its members access to weather and climate data within the Paso Robles AVA. During the mid-1990's, a team of meteorologists and physical scientists, working under the direction of Western Weather Group LLC, established a network of automated weather stations, which were installed in vineyards in the higher terrain of the Santa Lucia mountain range, northeast of the city of Paso Robles near the airport, in the far eastern growing region near Shandon, between the town of Templeton and Paso Robles west of Highway 101, and along Highway 46 in the Templeton Gap.

These automated weather stations provide real-time and historical weather data for the region (see photo to the right of an air and soil meterological station at Carmody McKnight vineyards). Members of the Paso Robles WCA can access the weather information and detailed weather forecasts. Today, these year-round weather forecasting services are used to support vineyard cultural practices. These automated weather stations provide the meteorological data used to make localized weather forecasts by supplying real-time and historical data to initialize and verify each forecast. Additionally, more automated weather stations have been established within private vineyards to characterize specific mesoclimates and microclimates in the area. Some of these automated weather stations are incorporated into the weather forecasting service that the Western Weather Group LLC team of meteorologists utilizes to produce daily Paso Robles AVA weather forecasts (Internet site: http://www.pasowine.com).

"IF GOD FORBADE DRINKING,
WOULD HE HAVE MADE WINE SO GOOD?"
CARDINAL RICHELEU

SAN MIGUEL DISTRICT (approximately 30,800 acres): warmer region III climate, with reduced maritime influence due to rain shadow of Santa Lucia Range, lower elevation footslope and valley floor topography, recent and Quaternary deposits of Salinas and Estrella Rivers predominate, with alluvial soils on floodplains and alluvial terraces, grassland with scattered oaks on hillslopes and trees along river terraces (Elliott-Fisk, 2007).

CAPARONE WINERY

What were you doing in 1973? We certainly were not writing in front of computers...but Dave Caparone was already producing wine from Central Coast grapes. He had loved wines from the classic European regions: Barolo, Bordeaux and Brunello, but was having a difficult time finding comparable wines in San Luis Obispo county where he had lived since 1966. In 1973, he made his first wine, a Dusi Ranch Zinfandel no less, and the horse was out of the barn. Inflamed by the marvelous results of this effort, Dave began studying winemaking in earnest, and soon growers were seeking him out for his talents. This allowed him to make wine from a variety of vineyards, soils and mesoclimates. He even put his native Italian-speaking father to work translating obscure Italian viticultural papers. His hobby getting completely out of hand, the seed was planted in his mind to start his own winery.

After more than half decade of meticulous research, he bought the site of Caparone Winery in 1978, on San Marcos Road just northwest of Paso Robles. His first wine under his own label was a Cabernet from the Tepusquet Vineyard in Sisquoc, but a long time goal of growing and making Nebbiolo simmered brightly on a back burner... After extensive research into the myriad, and at the time, utterly arcane world of Italian varietals, He got the first Nebbiolo vines in 1980, from UC Davis. Because of ongoing research in Italy and the US, they have since planted even more suitable clones.

Sangiovese Grosso (the Brunello clone of the grape) was planted in 1982, from cuttings from the famous "Il Poggione" Estate in Montalcino. The testament to the quality of those first harvests from those vines in the mid-1980's is that the wines are still aging gracefully more than 20 years after bottling.

Dave became interested in Aglianico, a grape with thousands of years of viticultural history, spanning pre Greco-Roman times, from its ancestral home in the Basilicata and Campania regions of southern Italy. It turned out to be more difficult than either Nebbiolo or Sangiovese to acquire. With the assistance of the venerable Dr. Harold Olmo of UC Davis (recently deceased), Dave finally got some cuttings from what turned out to be the last few true Aglianico vines in the US, and planted the first American vineyard devoted to this Aglianico varietal.

Vinification at Caparone is carried out in time-honored, low intervention manner. Dave's son, Marc, is his only partner in the process. Aside from the modern aides of temperature control, stainless steel and the like, all the wines undergo hand racking, extended skin contact and barrel aging, and are unfined and unfiltered. No micro-oxygenation or centrifuges here. The wines reflect a sense of place and varietal unlike many in California, or the world, for that matter.

A family heritage, a special location, vines selected and treated like family, and old world winemaking at its very best...this is primarily why Dave and Marc Caparone have succeeded. They are definitely worth seeking out; enjoy their wines!

CAPARONE WINERY
2280 San Marcos Road
Paso Robles, CA 93446
Phone: 805-467-3827
Fax: 805-238-9416
Web: www.caparone.com
Email: info@caparone.com
Hours: Daily 11 a.m. - 5 p.m.
Case production: 4,000

San Miguel
San Miguel, CA

Google Earth image (above): Aerial image of San Miguel and the surrounding vineyards and lands. Note the color differences indicating that the airphoto is a mosaic of images taken during the dry season (brown vegetation at left) and the wet season (green vegetation at right). Photo 16 (left): Aerial view to the east across the Salinas River and the Rinconada Fault line showing lands located immediately north of San Miguel, California (2 January 2001).

LION'S PEAK WINERY

Lion's Peak Winery came into being in 1992 as the vision of Ken and Jennifer Soni. Ken's modesty and mischievous smile disguises his raw determination that has marked all his successes. He originally came to America to escape the political turmoil in South Africa, landing in New York City with sixteen dollars. After working day and night to put himself through dental school, he eventually met and married the lovely Jennifer, a true soft-spoken Texas belle. They established a successful cosmetic dentistry business in Long Beach, and began visiting California's Central Coast on vacations.

In January 1991, over brunch at a local Paso Robles bistro, they found a real estate magazine with a picture of an old Victorian house on a producing vineyard. It seemed like a great spot to retire…lovely, remote, and, "they were tired of going to the same six vineyards…" After a Valentine's Day visit to the site in 1991, and subsequently making an offer, there were two hours of silence in the car on the way back to Southern California. "What if they accept the offer?"

Well, they did accept the offer, and while refurbishing the house, Ken made a barrel of 1992 Cabernet Sauvignon from his personal (own rooted, by the way…a real rarity) vines, to give away to friends. One of these friends submitted the wine to the Orange County Fair, where it promptly took a gold medal. Ken doubled the production to two barrels and took Best of Show in 1993. This same 40-acre rustic vineyard, originally planted in 1971 at 1,200 feet elevation, is still on its own roots. It was lovingly restored over the course of five years, after years of neglect, to its current thriving condition. Ken acquired more land, for 120 acres total; all are sustainably farmed, with little or no chemical usage… Ken still makes all the blends around his huge kitchen table and sends recipes he works out to all of the Lion's Peak Wine Club members.

The Flagship of the winery is, of course, is the estate grown Cabernet Sauvignon, and Cabernet Sauvignon Proprietors Reserve. Both are 100% Cabernet, and exhibit a rich, detailed, harmonious approach, spending three years in French and Hungarian oak. Another red blend of note is the Lionesse, a Cabernet, Merlot and Cabernet Franc blend, joined by a beautiful Paso Robles Zinfandel, a delightful lightly oaked Chardonnay, a lush, plumy Merlot, and a spectacular Viognier sourced from dry farmed Monterey Vineyards, which is also available in an unusual Late Harvest style. They are experimenting with Syrah, Sangiovese, Mourvedre, and Zinfandel, producing 20 varietals from the winery facility in Santa Maria. They are getting raves from the beautiful new tasting room in downtown Solvang. Ken and Jennifer have filled it with original oils, massive European wood furniture and antiques.

The name? Lion's Peak label sports a rendition of a 200-year-old painting, a birthday gift to Ken. Jennifer says the nose and attitude of one of the lions matched Ken's. Alternatively, it could be a reference to the multitude of wild game cats on the estate…You decide for yourself. Ken and Jennifer Soni's passion to make world-class Cabernet Sauvignon literally jumps out of the glass at you…the wines speak for themselves.

LION'S PEAK VINEYARD SALES & TASTING ROOM
7320 Cross Canyons Road
San Miguel, CA 93451
Phone: 805-467-2010
Fax: 805-467-3436
Web: www.lionspeakwine.com
Email: info@lionspeakwine.com
Case production: 1,290

LOCATELLI VINEYARDS & WINERY

So many times, we Americans erase our immigrant heritage from our memories. This now dimly lit pipe dream has been dwarfed by the new wave of hopefuls from Mexico, South America, Asia, and, indeed, the entire world. The whole earth continues to beat a path to our door. First, the religious persecutions in England drew Anglo-American countrymen and women to the east coast in the late 1600's...then, much to the dismay of their now naturalized, organized and frankly self righteous "natives" came millions of Germans, Irish, Swedes, Chinese, Italians, and many other nationalities...

They came here in the 1800's, then as now to escape grinding poverty, famine and disease, to try to make better lives for themselves and their families. How ironic we all share the myriad last names that reflect this diversity so deeply, and yet so many of us lose touch with our roots. Locatelli is a big exception.

Cesare Locatelli, the great-great grandfather/patriarch, dreamed of California from his ancestral homeland in Cerentino, Switzerland, near the border with Italy. At the age of 13, he journeyed to France, to work and save up the money for passage to the United States, and ended up borrowing a hundred dollars to fund the journey. He made it here, and worked at anything that came his way, to get to California.

He worked on ranches in Calaveras and Fresno counties, saving up to buy a property of his own, in the Fresno area, near Selma, California. After he built a house on the ranch, he went back to Italy to find and marry Aurelia, who was from a village named Bignasco. Upon their return, they farmed grapes, cotton, alfalfa, and operated a small dairy operation. They had six children, three boys and three girls. In another sadly typical hardship of the times, all three of his daughters died, including the last one in childbirth along with Aurelia.

He needed help with rearing the now motherless three boys, and sent for Assunta, Aurelia's sister in Bignasco, to assist him. In typical tough Italian peasant fashion, she made the passage to California alone, and this was long before jet travel, folks. Cesare taught his sons to farm, and crushed grapes every year, producing, and some say selling, wine even through Prohibition, but he never opened a winery. That would fall to his great grandson, Louis Gregory, and his wife, Raynette, who purchased the current property in San Miguel.

Louis and Raynette are ex-residents of Visalia...and bought 100 acres in 1996. With the help of Raynette's father, Ruben Gruber, they planted 40 acres of vines on the old cattle ranch. Ruben continued to manage the property until 2001, when Louis and Raynette moved out full time with their growing (five children, all under the age of 15) family. They grow Bordeaux varietals, Petite Sirah, Muscat Cannelli, and Zinfandel.

It is an entirely family run operation, with Louis making the wines in a true artisanal style, and Raynette running the tasting room and office. Louis makes an Estate Cabernet Sauvignon, Merlot, and Petite Sirah. The production is limited, all the reds receive lengthy

(2-3 years) aging in French and American oak. There is Aurelia, a sweet dessert style, and an unusual dessert style Cabernet Sauvignon. Cielo Rosso (Red Heaven...I love this name) is a blend of the Estate Cabernet Sauvignon and Estate Merlot. Luminoso is the name for their rose of Cabernet, and they produce a Viognier from purchased fruit.

It is great when the past isn't chewed up by the present and future; when families such as this, both thrive and stay true to their roots. More power to the extended family at Locatelli Vineyards & Winery!

LOCATELLI VINEYARDS & WINERY
8585 Cross Canyon Road
San Miguel, CA 93451
Phone: 805-467-0067
Fax: 805-467-0127
Web: www.locatelliwinery.com
Email: info@locatelliwinery.com
Hours: Friday-Sunday 11 a.m.-4 p.m.
Case production: 1,000

PRETTY-SMITH VINEYARDS & WINERY

Pretty-Smith Vineyards and Winery is a lovely, unusual jewel off the beaten path, nestled in the gently rolling hills at the northern edge of the Paso Robles AVA. Just two miles from the Mission San Miguel, Lisa Pretty-Smith's vision has its roots in her native home in Canada, where her father made wine at home...she still blames him to this day for "giving her the bug."

She is one of the very few female sole proprietors of an estate winery in the entire region, and justifiably proud of it. As the sole winemaker, vineyard and business manager, in her own words: "Lisa Pretty does it all for Pretty-Smith." Lisa bought the old Mission View Estate Winery in 2000, after leaving a very successful career in the information security business. Mission View is still the second label for the winery, offering gold medal winning value for fresh, fruit-focused wines...

All the Pretty-Smith wines are estate bottled, made from primarily Bordeaux varietals, grown on 45 acres of her 80 acre holdings. These are truly handcrafted wines, from 24-year-old vines, made from the classic Bordeaux varietals: Cabernet Sauvignon, Cabernet Franc, Merlot, and Malbec. Her signature wine is the Palette de Rouge: a delicious, warm, complex red blend of these varietals, and has won multiple awards. She also makes an Estate Bottled Merlot, Cabernet Franc, Cabernet Sauvignon, and a knockout Syrah...

Lastly, beautiful dessert wines from Zinfandel: a port style, as well as a late harvest style, both very limited and worth seeking out.

What's in the bottle isn't the only beautiful thing about Pretty-Smith wines...Ms. Pretty also designs the wine labels, basing them on her original paintings of ancient Southwest Indian symbols. They are unique and decidedly personal, adding another layer of individualism to this lovely property. Her fabulous standard poodle, Kokopelli, named after one of these symbols, is a real trickster in true Native American legend fashion. Lisa is currently producing less than 3,000 cases per year, and will limit production to 10,000 cases. Her large tasting room, decorated in a modern, colorful Southwest style, is a very befitting to the wines and spirit of the vineyards and rural location.

Pretty-Smith Winery is also a destination winery, sporting a large outdoor patio overlooking the vineyards. It is a fantastic place for a barbeque, winemaker dinner, reception or art show...there are picnic facilities available as well. The winery hosts monthly Wine Club events, wonderful twilight concerts in the fall, and participates regularly in the schedule organized by the Paso Robles Wine Country Alliance (PRWCA). These are truly delicious wines, a great story, and a beautiful place to visit.

PRETTY-SMITH WINERY & VINEYARDS
13350 N. River Road
San Miguel, CA 93451
Phone: 805-467-3104
Fax: 805-467-3719
Web: www.prettysmith.com
Email: info@prettysmith.com
Hours: Friday-Sunday 10 a.m.-5 p.m.

RABBIT RIDGE WINERY AND VINEYARDS

Erich Russell owns this beautiful, 55,000 square foot Mediterranean-inspired winery on the north side of Paso Robles. The winery name is a reference to his track and field days at San Diego High, when was known as the Rabbit. It is also a classic story of a home winemaker making it to the big time. The original Rabbit Ridge winery was in Sonoma County, opened in 1981, with two barrels of Zinfandel and two barrels of Chardonnay, made on the side while he was still teaching math and coaching track, before his leap of faith to the wine business. "I figured if I couldn't sell it I could drink it..." He made a name for himself after leaving Chateau St. Jean, where his summer job in the vineyards had morphed into a full time job. He still owns the original tasting room, and 40 acres there, but his heart (and production) is fully ensconced in Paso Robles. His early Sonoma wines attracted national attention, and attention for Sonoma, in general, mushroomed. Land prices skyrocketed, forcing many to inflate prices or yields, both unacceptable to him. Erich sought out a place to carve out another legacy, and Paso Robles, with its unique calcareous soils and a climate perfectly suited to viticulture, fit the bill perfectly. He began acquiring land on the westside in 1996, planting vines in three different vineyard sites in 1997.

As he points out, on the westside, " you get average temperatures five to eight degrees cooler than the Russian River..., mistakenly, everyone thinks Paso Robles is very hot..." but his firm, balanced wines are indicative of the coolness of the westside mesoclimate.

By 1999, he met and married Florida native Joanne James, his muse and partner. Joanne designs all the labels and marketing materials, assists with accounting, and has written a great cookbook, "A Cook's Tour of Rabbit Ridge Winery," with recipes collected and perfected over the last 15 years. The design for this high tech, gravity flow winery began in 1996; construction was completed in 2001, in time for the 2002 crush.

They own close to 700 acres, spread over several westside vineyard sites: Cristalla Ranch, Vista Serrano, Texas Road, and the Chimney Rock property, at 2,000 feet elevation and about 10 miles from the Pacific Ocean. They grow 20 varietals spanning the gamut of Mediterranean choices: Zinfandel, Primitivo, Petite Sirah, Barbera, Dolcetto, Brunello Clone Sangiovese, Refosco, Viognier,

Rhone varietals, and Bordeaux varietals. All the vineyards are planted in a very closely spaced pattern, and yields are very low, around two tons per acre. Tuscan olive trees round out the picture, interplanted with lavender.

Needless to say, with this many varietals, you could fill a small tome with all of these interesting, beautifully crafted wines. A beam of balance and richness runs through all of them. Erich's use of oak is restrained, serving to frame rather than overwhelm the fantastic fruit flavors. His winery is a melding of science and art, with computer-monitored tanks, automated pump-over regimes that Erich can access from his home computer, if necessary.

There are a number of low production unique blends in the picture: a Cote Rotie style Syrah, blended with Viognier, and Vortex, a Bordeaux/Rhone blend. There is a Brunello clone Sangiovese, an unoaked Viognier and similarly styled Chardonnay, along with the prerequisite Reserve Zinfandels, Cabernets and Petite Sirah. His answer to the Primitivo vs. Zinfandel controversy is succinct: "I used to say they are exactly the same, but they're not. We're one of the only wineries in the world where you can taste them side by side...Primitivo is like Zinfandel on steroids...richer and little bit more jammy, a little less raisiny or pruney, at the same alcohol levels. Just a wonderful grape. They both need to be really, really ripe to make good wine....Primitivo bunches ripen more consistently, solid clusters of evenly ripe fruit."

This is in contrast to Zinfandel, which will have raisined grapes throughout its clusters, at the same Brix levels. These wines go through extended maceration, 30 days on the skins. They use as special alcohol tolerant yeast from the Cote Rotie, pumped over half a dozen times a day. The regime is similar for all of his red wines, except for Pinot Noir, with a much more limited weeklong maceration.

Erich designed the entire winery geared to production: wider tanks that ensure a thinner cap, state of the art pump over systems, gravity flow. It is an elegant, efficient marvel. "I don't like factories, and I wanted this to be a nice place to work..." Water issues are addressed with multiple reservoirs, conservation of winter rainwater, and reuse of winery waste water to irrigate the vineyards.

They also produce a Central Coast Appellation Barrel Cuvee series, and a multi-appellation specialty series. They are on their second release of olive oil pressed from the Tuscany varieties. Erich Russell's dream of two plus decades has come to fruition in fine fashion, and will no doubt be a benchmark Paso Robles success story in years to come.

RABBIT RIDGE WINERY AND VINEYARDS
1172 San Marcos Road
Paso Robles, CA 93446
Phone: 805-467-3331
Fax: 805-467-3339
Web: www.rabbitridgewinery.com
Hours: Daily 11 a.m. - 5 p.m.

RAINBOW'S END VINEYARD & WINERY

I drove up to Jim and Shirley Gibbons' property one of those perfectly clear, hot, summer Paso mornings. Their home, surrounded by vineyards, sits atop a hill, basking in the sun, very Tara-like. An enormous weeping willow tree (almost) overshadows the large, pillared veranda, and under it, distinctly un-Tara like, was parked a fantastic chopped, Day-Glo orange and blue flame Harley. Add to this the roaming peacock, Joe, and you have it; a classic scene from Paso Robles, the old with the new, the shiny with the sublime.

Originally, in the underground water pipeline business in Los Angeles, the Gibbons came here looking for a retirement site, their rainbow's end, if you like… They purchased the property in 1980, and spent three years building their wonderful house. Seduced by the wines in the area, Jim began to study viticulture, and produced several award-winning wines in the early years. As the potential of the site became clearer, they decided to open a family owned and operated boutique winery, specializing in small lots of handcrafted wines.

It is a 35-acre site, with 25 acres planted to Merlot and Cabernet Sauvignon, and future plantings planned for Syrah, Barbera and Petite Sirah. Production is still small, around 2,800 cases, and as Jim laughingly says, "there's a lot more money in water than in wine." This seems prophetic with what is happening as water yields are reduced in water wells located in the vineyard hills west of San Miguel.

They are planning to purchase property on the west side of Paso Robles, and grow some cooler climate varietals, but Jim's favorite wine to make is Cabernet Sauvignon. It is too warm here for his favorite varietal, Pinot Noir. He is often in the vineyards, and the hands-on vineyard management pays off in superb fruit quality. Their son, Chris, has joined the winemaking team, having produced his first vintage in 2003. All the Cabernet Sauvignon and Merlot is estate bottled, made in a warm, sunny, round style, with excellent fruit character and good balance. They spend enough time on the skins, and in French and American oak, for aromas, extraction, and color to evolve…but are not hard wines. Lush and forward would be more descriptive terms.

They make a number of other wines: Zinfandel, Nebbiolo, two Syrahs: from the Cagliero vineyard, and San Marcos Creek vineyard in a somewhat firmer style, a lush Paso Robles Chardonnay. There is Petite Sirah, Barbera, and the latest addition to the portfolio is Prism: a fantastic Merlot and Cabernet Sauvignon blend that is already winning awards. The tasting room is a small, cool, inviting boite, open most weekends or by appointment. The property is a great site for special events, as well. The Gibbons are a warm, generous couple…and their son, Chris, will build you a fantastic chopper or custom rod, but the real story here are the wines. Jim and Shirley Gibbons are making their dream of "wines that people will remember" come true. A visit to Rainbow's End is a must for all wine lovers!

RAINBOW'S END VINEYARD & WINERY
8535 Mission Lane
San Miguel, CA 93451
Phone: 805-467-0044
Fax: 805-467-2304
Web: rainbowsendvinyard.com
Email: rainbow@digitalputty.com
Hours: Friday-Sunday 11 a.m.-4 p.m.
Production: 1,500 cases

RIVER STAR VINEYARDS & CELLARS

Uh-oh, here we go again. A simple retirement idea gone awry...after a previous life as ranchers in the Paso Robles area, Ed and Muriel Dutton were just looking for a place to retire from their Ontario, California paper company, when they came upon some land near San Miguel. The property had a few untidy Cabernet Sauvignon vines on it. That was in 1986, and with some help from friends the vineyard soon shaped up, so well that they got a lovely harvest in the banner year of 1987. That spelled the end of retirement, and RiverStar vineyard was launched in 1993. In 1997 they bought another parcel planted to Cabernet Sauvignon and Sauvignon Blanc. They sold their superb fruit to the likes of Meridian, Kendall Jackson and Treana, before releasing their own wine starting in 2001.

The vineyard team was mentored by Don Ackerman, and practices sustainable farming in every way possible. Winemaker (and son-in-law) Michael Coyne was taken under the wings of consultant Dan Kleck at the Paso Robles Wine Services, where they are currently making their wine, utilizing 95% estate grown fruit. Michael makes a Sauvignon Blanc, aged in neutral French oak, but his red program utilizes about 30-35% new French and American oak. There is a Merlot blended with a little Cabernet, a Cabernet Sauvignon, and a Syrah that sees 100% French oak barrels for over a year.

They release "Affinity," a 40% Cabernet, 40% Merlot, 20% Zinfandel blend, and have commandeered some of Don Ackerman's own Petite Sirah to release as well. Michael has found a source for Touriga Nacional, one of the original Portuguese Port varietals, and has crafted a port style wine. They also are planning to release some red Rhone blends...

Perhaps most exciting, the new winery opening near Airport Road, which will enable them to put "Estate Bottled" on their wines, an onsite tasting room, and give us all another reason to visit and enjoy Paso Robles.

RIVER STAR VINEYARDS & CELLARS
7450 Estrella Road
San Miguel, CA. 93451
Phone: 805-467-0086
Fax: 805-467-2846
Web: www.riverstarvineyards.com
Email: greatwines@riverstarvineyards.com
Hours: Thurs.-Mon., 11 a.m. - 5 p.m.
Case production: 1,500

Directly off the 101 freeway, just north of Paso Robles, a brand new dream is taking shape. San Marcos Creek Winery is the vision of Catherine and Brady Winter...the tasting room and Bed and Breakfast have been completed since my initial visit. Now that it is finished, Paso's newest destination winery will be a wonderful combination of Paso charm and old European style. The Tudor-influenced architecture and warm colors are a fabulous combination and the genuine open, friendly way that Catherine and Brady add to the appeal of this winery.

The Winters are partners with Catherine's parents, Fling and Annette Traylor (who were high school sweethearts, no less). Fling founded the Hose King Company in the City of Commerce, and was very successful for thirty years. In 1981, while they were searching for a retirement home in the Paso area, they happened upon a home on an old ninety-acre sheep ranch, which they purchased. Not exactly as successful with sheep as with industrial hoses, they brought in horses, and leased out their land.

The Traylor's could not help but notice a vineyard-planting boom happening all around them. Their property, with its rich mineral soils and deep Salinas aquifer water, was ideally suited for grape production. The first plantings of Cabernet Sauvignon, Syrah, Nebbiolo and Merlot went in 1990, with a majority being Cabernet. They had immediate buyers for their first commercial crops, and moved to Paso full time in 1994. After closing Hose King in 1996, new construction started on the property. They initially intended to be a "gentlemen farmer-vineyard-Bed and Breakfast" operation.

In the late 1990's, though, a grape glut (which periodically still occurs) encouraged them to build a winery and launch a wine brand. Catherine Winter was a golf course business manager for 19 years, husband Brady was in the automotive finance sector for 15 years. With an 11-year-old son in tow, they made the move to Paso Robles in 2002. Brady helped finish the winery, recreated from the old sheep barns, and opened for business in November 2003.

With a 7,000 case potential, they have produced 2,000 cases in 2004. Wine consultant Paul Ayres of Paso Wine Services is assisting Brady with winemaking, and the assistant winemaker is Robert Hall, of Castoro Cellars. The new releases are full, rich wines...perfectly balanced and worth seeking out, brought up in mostly neutral and 25% new French oak. Brady's favorites to grow and make are Zinfandel and Syrah...they grow no white grapes, but source a fine Viognier from Dunnigan Hills in Yolo County. All the wines except this one are estate bottled, and most of them are from single vineyards as well...

There is a Reserve Cabernet in the pipeline, crafted in a more extracted style...extended cold soak maceration and two years in French oak barrels. The wine is showing real promise. Brady is doing some interesting experimentation with hybrid oak barrels; barrels with French oak heads and American oak staves. He is looking to balance out the somewhat overbearing aromatics of American oak with a more subtle French influence. The wines are minimally filtered before bottling. These are truly appealing, fruit focused wines with excellent structure and balance.

They have a small lake on property, a fabulous new tasting room, and a bed and breakfast inn to match. This is a winery to be visited and closely watched.

SAN MARCOS CREEK WINERY
7750 North Highway 101
Paso Robles, CA 93446
Phone: 1-866-PASO-WINE
Fax: 805-467-0160
Web: www.sanmarcoscreekvineyard.com
Hours: Daily 11 a.m.-5 p.m.
Case production: 2,000

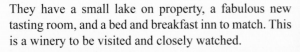

SILVER HORSE VINEYARD & WINERY

There is a gorgeous new Hacienda style winery and tasting room perched on top of a hill east of San Miguel. The Kroener family started the vineyard (an 80-acre property, of which 40 are currently planted to vines) in 1989. The winery was bonded in 1991. The beautiful new tasting room and winery was re-opened for serving in May of 2005.

Steve Kroener's grandfather made port and dessert style wines by hand in Los Angeles (yes, there was a time when LA was mostly orange groves and vineyards) and managed to infect the young scion with the winemaking bug. Besides his grandpa, he lists such influences as Justin Smith of Saxum, with his decades of vineyard experience, and Matt Trevisan of Linne Calodo "...we're all the same age, and just hung out, picking up knowledge and techniques..." He also took courses at UC Davis and Cal Poly, but is largely self-taught.

The vineyards are planted to Cabernet Sauvignon (which is own-rooted) Malbec, Petite Verdot, Tempranillo, Garnacha (Grenache) and Syrah. Soon to be planted are Albarino, Tannat (a hugely tannic, very dark, brooding, red varietal that originated in the southwest of France), Petite Sirah and Zinfandel. They are developing sustainable vineyard practices on an ongoing basis, and will be adding solar power to the winery.

Steve uses a varied oak barrel pallet: French, American, Hungarian and Russian "...I love the quick extraction of flavors from the Russian barrels...and the meaty, bacon-fat complexity that the Hungarian oak adds to Syrah...they are very tight grained, not unlike the French oak barrels." Most of the reds see a cold soak of 4-7 days and anywhere from 18 to 24 months in barrel. All but the Albarino (a very fashionable white varietal that originates in northern Spain) from a tiny Edna Valley vineyard, are estate bottled. He sells about 85 percent of his fruit to a very select group of wineries.

They produce varietal Tempranillo, Cabernet Sauvignon, Syrah, and some unique blends. "The Big Easy," a Tempranillo/Grenache/Cabernet Sauvignon/Syrah blend, also in a rose style, and "Sage," named after his elders' sage advice, is Bordeaux styled, based on Cabernet Sauvignon, Merlot and Malbec. They produce a varietal Albarino, and "Carame", named with his wife in mind, is also Bordeaux styled, consisting of Cabernet Sauvignon and Petite Verdot. "Tomoto", named for an amalgam of the children's initials, is a Syrah/Cabernet Sauvignon blend. "Tori," is Syrah/Tempranillo...he produces a Zinfandel port-styled wine named "Sozinho" as well. All these lots are less than 200 cases. When asked what drove him to Paso Robles, Steve is quick to answer, "A love of vino!..." and this is what anyone that visits their vision in San Miguel will come away with as well.

SILVER HORSE WINERY
2995 Pleasant Road
San Miguel, CA 93451
Phone: 805-467-WINE
Fax: 805-467-9414
Web: www.silverhorse.com
Email: steve@silverhorse.com
Hours: Friday-Monday 11 a.m.-5 p.m.
Case production: 500

VISTA DEL REY VINEYARDS

Have you ever gone on a back roads driving tour? The trip to Vista Del Rey Vineyards has that feel to it. Surrounded by old west, cattle and horse ranches, several miles west of the 101, only four miles south of the Monterey County line, lies the truly unique vision of Dave and Carol King. Up and over gravel roads, at the top of a big hill, one comes upon the Dave and Carol King's dry farmed, gnarled Zinfandel and Barbera vines. There is a big water reservoir tank with Vista Del Rey emblazoned on it, giving the only concrete hint of this vineyard's name.

Dave and Carol warmly welcomed me into their world, and as I stood on top of the hill overlooking the vineyards and valley floor below, I felt a sense of place, a real feeling of *terroir* that was unmistakable. This is one of the warmest areas in the Paso Robles AVA. Located on the eastern slope of the Santa Lucia Mountains, which acts as a marine layer barrier, the climate is nearly continental, with very hot, dry summers and cold winters. Yearly rainfall is normally about twelve to thirteen inches, significantly less than other mesoclimates in the Paso Robles AVA. Dave dry farms most of his vines, an anomaly for the area as well. The silty clay loam soils helps sore moisture in the vineyard, but the vines are stressed and much smaller than usual. His yields are correspondingly miniscule, less than a ton per acre.

Ushered back into the quiet, eclectic tasting room, Dave (along with dogs Shiloh and Daisy) elaborated at length about his wines and his story. He was 22 years in the Navy, did five tours on ships, including a tour in Vietnam. After postgraduate studies in Oceanography at the U.S. Naval Academy in Monterey, he taught for three years. He also completed two tours at high tech naval labs. His rather significant accomplishments include helping to design the current system for undersea surveillance, as well as weapons. This in turn led to a career in high tech undersea research. The end of the Cold War in the early 1980's has (mostly) meant the end of this type of activity, and freed Dave up to focus more on his "Walter Mitty dream" as he puts it.

He was interested in moving to the Central Coast area for some time. With a love of the beach (he is an ex-two-man volleyball aficionado), as well as writing an independent wine newsletter "King's Corner of the Wine World." He was moving in the direction of Paso Robles…his research took him all over California, including a stint during crush at Wild Horse, finally deciding in 1994 to buy eight acres of an old barley ranch. He owns 22 total acres now, almost all dry-farmed Zinfandel and Barbera. These are not your typical Paso Robles wines in any sense due to the intense flavors originating from Dave's dry-farmed vines.

These wines have a firm acidity, very balanced extract, an almost European feel to them. Not your fruit-jammy. sweet wines at all. Dave and consulting winemaker Larry Roberts have done an excellent job coaxing the fruit out while maintaining balance. After destemming and crushing, the fruit is fermented in open top vessels, hand punched down twice a day…the wines spend close to a year in passive and 20% new French and American oak. There are several singular wines: the "El Zino" 1998 from the relatively wet El Nino vintage, is one of the best I've had from this difficult year…there's a "Mundo Viejo" ("Old World") Zinfandel. He makes a fabulous "Toro Negro" Barbera (Black Bull, named after a rogue black bull that made a home in the vineyard one year), 100 percent Barbera, from the nearby four acre Colina de Robles Vineyard. They also make a very limited rose of Zinfandel. Uncommonly, he has multi-year verticals available as well... an outstanding opportunity to taste these wines as they evolve.

In addition to the wines, the tasting room has a plethora of gourmet food items, including handmade chocolate, some of these made with Vista Del Rey wines, and a number of interesting displays, including a history of the Zinfandel grape. Dave and Carol King are living their dream, and making wonderful wines to boot. Word of mouth has spread that they are gracious hosts, and although available for tours and tasting by appointment only, but it's well worth the extra effort. As Dave puts it, "people come here and we're not going to be like bumps on a pickle…we treat people like family." Go see for yourself…

VISTA DEL REY VINEYARDS
7340 Drake Road
Paso Robles, CA 93446
Phone: 805-467-2138
Fax: 805-467-2765
Hours: Most Sundays and special event weekends;
11 a.m.-5 p.m. and by appointment.
Case production: 700

CHAPTER FOUR: ADELAIDA DISTRICT

ADELAIDA DISTRICT (approximately 53,100 acres): cooler region II-III transitional mountain climate, with modest maritime influence and cold air drainage downslope, high mountain slopes grading to base of foothills, bedrock residual soils and colluvial soils from middle member of Monterey Formation and other geologic formations, variable depth soils are mainly calcareous with some siliceous shale beds, with oak woodlands to mixed woodlands in hills and mountains (Elliott-Fisk, 2007).

ADELAIDA CELLARS

Any discussion of this distinctive Paso Robles mesoclimate should begin with nod to the historic property, beautifully situated on a hilltop ranch, with elevations to 2,300 feet. During the 1870's, the Adelaida district was a busy hub of activity, with six to seven hundred souls served by a full complement of social services. The middle Salinas Valley needed an outlet to the Pacific Ocean, supplied by the packet steamers that plugged up and down the coast. Once the Southern Pacific railroad reached San Miguel and Paso Robles, that Pacific access was no longer required...the inevitable decline ensued, but the Adelaida area has retained its distinctive identity.

The unusual, highly calcareous shale and mudstone-derived soils drew Ignace Padrewski, the famed pianist and Polish ambassador, who planted Zinfandel and Petite Sirah in the 1920's at his Rancho San Ignacio. Dr. Stanley Hoffman, who planted several hundred acres of Pinot Noir in 1963, continued the legacy. The HMR (Hoffman Mountain Ranch) is one of the oldest Pinot Noir vineyards in the state of California.

The Don and Elizabeth Van Steenwyk family purchased Adelaida Cellars in 1991 from founder John Munch, and three years later, obtained a 400-acre parcel that included the original HMR vineyard. They have revitalized the vineyards, and continue to develop the plantings. In addition to their HMR property, the Viking Estate Vineyard is very suitable for Cabernet Sauvignon. Located 16 miles from the Pacific coast and near the southern end of the Santa Lucia Mountains, the dramatic diurnal temperature fluctuations have a profound influence on the grapes, as does the 25 to 50 inches of annual rainfall, and the general lack of coastal fog. There is also the Bobcat Crossing, planted to Portuguese varietals: Touriga Nacional, Tinta Cao, and Souzao, as well as Muscat Blanc. They also make their own wood-aged brandy, for use in fortifying their port styled wines.

Eighteen acres of Syrah are grown on the ridgetops contiguous to HMR. A Rhone planting has been completed, with Beaucastel clones of Grenache & Mourvedre. Petite Sirah from this vineyard will be added to their Zinfandel. There is also a new Heritage Clone Zinfandel planting (cuttings from the Will Pete Vineyard, planted in the late 19th Century). These vines will be dry farmed and head trained. The vineyard practices are committed to sustainability, with a no-till policy that encourages native grasses, wild sage and rosemary between the vine rows.

COMPETING NEIGHBORS

Mediterranean ecotypes in California are being
Replaced by crops species from comparable
French, Italian and Arabian origins.

Will these foreign cultivars aid or
Harm their aboriginal predecessors?
How can we fully understand this complexity?

Deer and turkey gorge on the grape leaves and fruit.
Rabbits and squirrels relish the rich walnut.
Fences of electricity built to deter the hungry pests.

Deep ancient aquifer water is discovered and
Pumped to the surface to supply the
Fluid refreshments to thirsty grape vines.

Is this ancestral water; are the deep aquifers
Recharged following each winter rainfall?
Does anyone really want to know?

Settlement, utilization and competition;
All inevitable consequences when our
Human appetites and population numbers increase.

Photo 17: View from HMR Vineyard looking southwest across Peachy Canyon Road to walnuts, oak woodlands and the Bailey Vineyards (above: June 2001).

Photos 18-19: Backhoe excavating soil on potential vineyard lands located north of the HMR Vineyard (May 2001). Paul Sowerby lecturing Cal Poly students about Adelaida soils and wines (below: 16 May 2002).

Winemaker Terry Culton brings his past experiences with Pinot Noir legend Josh Jensen at Calera, as well as knowledge from Ken Volk at Wild Horse, and work at Edmeades. His philosophy of judicious use of barrels (small oak cooperage for the majority of the wines) and extended lees aging is adding an element of complexity and generosity to the wines. The Van Steenwyk family, a unique combination of talent and intellect, has succeeded in their quest to produce wines that truly reflect to potential of the property. This winery story is bound for continued future excellence.

ADELAIDA CELLARS
Address: 5805 Adelaida Road
Paso Robles, CA. 93446
Phone: 805-239-8980
Fax: 805-239-4671
Web: www.adelaida.com
Email: wines@adelaida.com
Hours: Daily 10 a.m.-5 p.m.
Case Production: 15,000

CARMODY MCKNIGHT ESTATE WINES

This is an endearing American story...well-known actor and artist turned vigneron, married to Miss America, emerges unscathed from his crashed helicopter to exclaim to an understandably distraught realtor, "I'm gonna buy this place!" It would be novel-worthy if it were not true. These days, it is script worthy, because it's true, should I say...to add to the uniqueness of the scenario, where this is the one of the few vineyards in the Paso Robles AVA possessing both volcanic (igneous) and limestone (calcareous) soils.

In 1989, Gary and Marian Conway founded the original winery, Silver Canyon; following the initial 1985 vineyard planting. Carmody McKnight became the winery name in 1995, with their estate bottled blends of Bordeaux varietals, along with three Chardonnay variations and Pinot Noir, grafted from the famous HMR cuttings. They also make a unique dessert wine, Kathleen, made from late harvest Cabernet Franc. More on the vines, in a bit...now to the soils...

Apparently, the shallow ocean sediments of the Monterey Formation were dissected by an intrusion of igneous rocks; causing the sedimentary marine deposits to be inset by extruding magma. Today, this vineyard is home to a unique combination of the igneous rocks and calcareous sedimentary rocks, including limestone, that weathers to montmorillonite (a clay mineral abundant in ocean sediments and called "nano-clay" due to its small size). What this means is that besides rarely fertilizing the vines, which would be enough to get the attention of anyone who grows plants, the fruit quality is outstanding and in high demand by many winemakers. These vineyards are also the location for long-term research studies by faculty from Cal Poly, Earth & Soil Sciences Department.

Winemakers Kathleen Conway and Greg Cropper have some remarkable natural resources to draw upon...Kathleen grew up on the property, with all the intimate vineyard and *terroir* knowledge that brings. Greg joined the team over ten years ago, moving from cellar master to vineyard manager and now to general manager and co-winemaker.

The wines they produce reflect the vineyard *terroir* in an intimate way...a Pinot Noir, aged in French oak. Cadenza, a Bordeaux blend, that garners gold medals from prestigious competitions; awards given not only for their wines, but also for Gary's colorful paintings that grace the wine labels. There is an estate Cabernet Franc, highly unusual in that it is 100% varietal, along with a Cabernet Sauvignon and a Merlot, also 100% varietal. The white wines are the antithesis of the oakey, overblown styles of California past. Aptly named "A Day in the Country" is 100% estate grown Chardonnay, 90+% stainless steel; and a "Free Run" Chardonnay is made from free-run juice (extract that runs out of the fruit before pressing, which results in clear, crisp fruit flavors), and touches a little more French and American oak. There is also a limited production of a single vineyard white wine, "Marian's Vineyard" Chardonnay, as a third variation. There is a fantastic selection of dessert wines as well; a late harvest Cabernet Franc, a fortified port style Merlot, a late harvest Chardonnay that is just sweet enough...and if this isn't enough, a sparkling Kathleen's Brut Cuvee made of the classic Champenoise varietals, with a little Pinot Blanc. Visit them each March to participate in the "Blessing of the Vines!"

This is a truly magical story...complete with olive groves that produce gourmet oils...it's a fabulous Paso Robles saga, on all levels. Additional in-depth information on the soils, landscapes, and vineyards of Carmody McKnight is discussed later (see Appendix A). Cheers to the fine wines and these magical vineyards!

CARMODY McKNIGHT ESTATE WINES
11240 Chimney Rock Road
Paso Robles, CA. 93446
Phone: 805-238-9392
Fax: 805-238-9375
Web: www.carmodymcknight.com
Email: info@carmodymcknight.com
Hours: Daily 10 a.m.- 5 p.m.
Case production: 3,000

HALTER RANCH VINEYARD

This is a relatively new winery with some very old roots in Paso Robles history. On a ancient Indian settlement, complete with an ancient spring, the main ranch house is set back in a gorgeous tree lined, park like setting. This meticulously restored Victorian farmhouse was once owned by Paso pioneer, Edwin Smith. Smith enjoyed remarkable fortune, building a grand house, dealing in livestock, farm produce, and kept a thoroughbred stable at the Estate. The house actually burned down but he promptly rebuilt it. At that point in history, it was part of his 3,600-acre estate, which was broken up in the early 1900's, after Smith's fortunes faded. In 1943, 1,200 acres were purchased by the MacGillivray family. The MacGillivrays farmed this land for 57 years, and introduced grapevines in 1996. The house is a Paso Robles landmark to this day...and a recent bit of history is its appearance in the movie "Arachnophobia."

In 2000, Swiss entrepreneur Hansjorg Wyss entered the story. Wyss is President of Synthes USA, a global leader in the medical technology industry, and an ardent environmentalist. He journeyed to Paso Robles, was immediately smitten, and purchased 900 acres of the original property from the MacGillivray family. The folks at Halter Ranch have spent the last six years improving the 250-acre vineyard, renovating the historic farmhouse, building a state of the art winery, and planning the construction of a larger tasting room.

They have planted 15 different grape varieties-60% Bordeaux, 40% Rhone, and including some own-rooted Syrah and Zinfandel. There are 42 separate vineyard blocks primarily based on soil types and slope aspects, ranging from soils derived from calcareous shale and mudstone rocks, to deep alluvial calcareous clay loam and clay soils on the Las Tablas Creek terraces. The elevation (up to 1,800 feet), steep slopes, and south-facing hillsides, add to their unique *terroir*, further enhanced by high-density plantings, low yields, multiple clones and rootstocks, and a deep commitment to agricultural sustainability. The ultimate goal is organic vineyard certification, and they are well on their way. Cover crops, in-row tillers, integrated pest management, all certified organic products used in the vineyard and winery...this is a vineyard and winery team totally dedicated to producing world class wines that genuinely reflect Paso Robles *terroir*.

Vineyard manager, Mitch Wyss, an avid outdoorsman and veteran of farm management, is producing some of the best fruit in the Paso

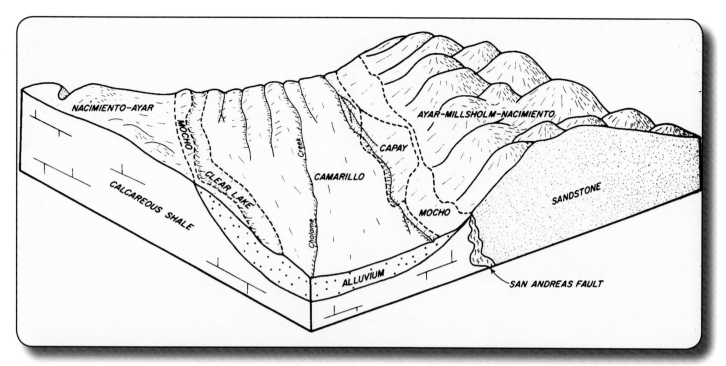

Figure 13: Block diagram showing geology, soil and landscape relationships in the Paso Robles AVA (USDA, 1983).

Photos 19-22: Ayar (Vertisol) soil planted with Cabernet Sauvignon on Tablas Creek terrace (above) and Nacimiento (Mollisol) soil planted with Merlot on hillslopes (below) located within the Halter Ranch Vineyard (16 August 2002).

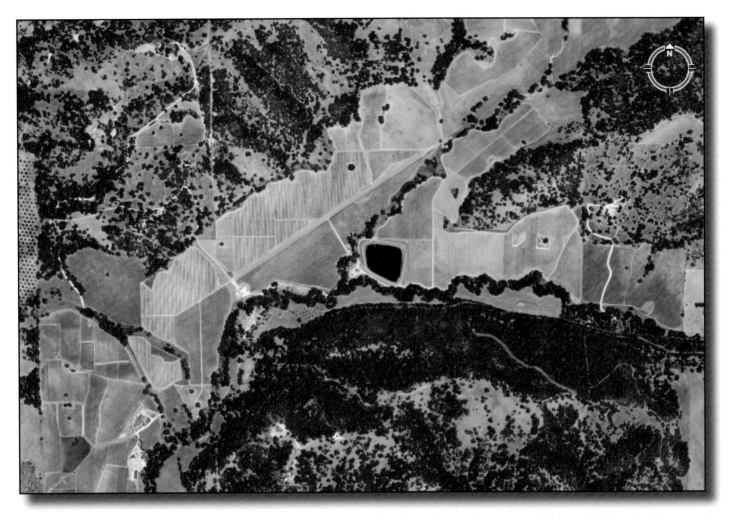

Figure 14-15: Google Earth image of Halter Ranch Vineyard and Tablas Creek Vineyard (above). Soil map of Halter Ranch (below).

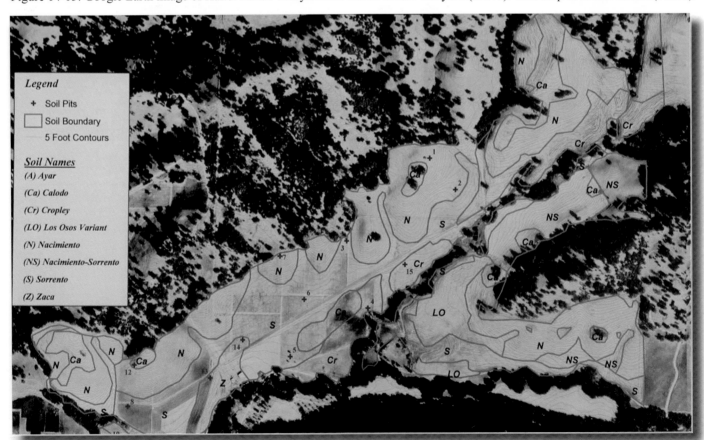

Robles AVA. Mitch is producing high quality wine grapes within vineyards that have a diversity of soil types representative of the soils in the larger Adelaida Hills region (see Figures 13-15 and Photos 19-22). Steve Glossner, a highly regarded veteran Paso Robles winemaker, came from Adelaida Cellars to produce delicious 2002 vintage reds, released in the spring of 2005. All the wines are fermented in small open top and closed top fermenters, with well-balanced extraction from punch downs and pump overs. His judicious use of French oak, 30-50% new, rounds out the wines' flavors. These wines are delicious and also improve with some cellaring. Steve retired from Halter Ranch in 2005 to teach enology at Cal Poly, San Luis Obispo, and to engage in private enology consulting. The Halter team immediately endeavored to hire another outstanding winemaker.

Bill Sheffer, Halter Ranch's winemaker, has over 20 years' experience making wine in the Paso Robles region, as well as throughout some leading wine regions throughout the world. Bill began his career at Estrella River Winery in the early 1980's, he moved on to assistant winemaker positions at Adelaida Cellars and J. Lohr Winery, before serving for eight years as head winemaker at Eberle Winery. Bill gained invaluable international experience at well known Southern Hemisphere wineries such as Viña Santa Rita in Chile and Rapara Road Vintners in New Zealand, along with a short stint at La Viña Winery in Valencia, Spain.

Halter Ranch is a great new addition to the Paso Robles AVA; complete commitment to quality, respect for the environment, and a long-term vision that is rapidly emerging on the wine industry radar screen.

Halter Ranch Vineyard and Winery
8910 Adelaida Road
Paso Robles, CA 93446
Phone: 805-226-9455
Fax: 805-226-9668
Web: www.halterranch.com
Email: info@halterranch.com
Hours: Friday-Monday, 11 a.m.- 5 p.m.
Case production 5,000

JUSTIN Vineyards & Winery

Sometimes a story just sparkles. It flows seamlessly and energetically from beginning to end, always interesting, even visionary. This is the case with Justin, who with a few other wineries has captured the wine loving public way beyond the borders of Paso Robles. A little lost on some folks is the fact that the proprietors have been at it for 25 years...

Justin and Deborah Baldwin wanted a real life, not just success. Deborah's background in rural Washington state, and Justin's Midwest (Pennsylvania and Minnesota) farming heritage combined forces that drew them naturally to wine country. Justin was raised in San Francisco, where his success as an international investment banker led him to visit the Napa Valley while entertaining clients. This lifestyle had always appealed to him...and he and Deborah, a soon to be ex-mortgage banker, bought 160 acres in the Adelaida Hills fifteen miles west of Paso Robles, in 1981. There were eight bonded wineries in Paso Robles then. Now there are nearly 200 wineries and the number grows annually.

They immediately planted Bordeaux varietals; Cabernet Sauvignon and Franc, and Merlot, Malbec and Petite Verdot. There are about 72 planted estate acres, including some delightfully orange flower scented Orange Muscat. They also grow Syrah, Roussanne, Tempranillo, and, yes, Zinfandel. The winery was built in 1985, and by 1987, most of the wines were estate bottled.

Director of winemaking is Fred Holloway, a Fresno State enology graduate, and veteran of over 20 years making wines, from hands-on to supervising over 800 people. He cut his teeth at wineries like Lokoya, Cardinale, La Crema, Verite, Edmeades and Cambria... and was getting the attention of world wine critics long before his tenure at Justin began in 2003. Kevin Sass, the winemaker, is a Southern California native and Fresno Enology and agricultural business graduate. He joined Justin as an intern for the 2000 harvest, was promoted in 2002 to Associate Winemaker as well as grower relations manager.

The focus in the vineyards is getting the fruit to ripen to the levels that Holloway and Sass feel are correct, not just by the numbers. They hone in on physical maturity, of skins, seeds and juice, not just chemical maturity. Their growers, and their estate vines, are meticulously managed...complicated canopy management, green harvest, and strict bunch triage before and during harvest all play a part. The growers are all invited every year to a blind tasting of the wines from their fruit, with ranking done before they know who

produced what. That alone is going to get the attention of any serious vigneron. Fred says he does not do most of the talking at these events, either. The vineyard growers are constantly piping up about how they can grow better fruit, and the results are in the bottle.

Winemaking here, is focused on framing the fantastic fruit...the reds get cold soaking, extended maceration, new French oak barrels, periodic deletage, little or no fining or filtration, and additional bottle age prior to release. They make a reserve Cabernet Sauvignon, the internationally famous Isosceles and Justification bottlings, both Bordeaux varietal based. These wines are sitting at the pinnacle of Paso Robles quality, and are deservedly scarce. The Cabernet Sauvignon sells out quickly, and they make a fantastic Syrah as well. They make a reserve Tempranillo, a grape with a real future here given its proclivity for heat. There is a Mourvedre rose, beautifully rich Chardonnay, and Sauvignon Blanc along with a White Rhone blend to round out the white wine program.

Their dessert wines have gotten a lot of attention; Obtuse, a (fortified) Port style wine is based on Cabernet Sauvignon with the addition of the classic Portuguese varieties. They also produce Deborah's Delight, from Orange Muscat, is almost like a Muscat de Baumes de Venise; fortified and then aged in stainless for a few months before release in a gorgeous package. The constant critical acclaim from all corners of the wine world, and the continual sellout status of the releases, are proof positive that they have hit the bulls-eye with every wine they produce.

If all this success and adulation were not enough, they have an excellent "auberge" style inn, the Just Inn, adjacent to the vineyards. Their restaurant, Deborah's Room, is headed by executive chef Ryan Swarthout, formerly of McCormick and Kuleto's, and the Carnelian Room. His hero is Thomas Keller of French Laundry. He has done an amazing job of bringing his philosophy of intricate preparations to the Paso Robles dining scene, including a deep respect for the integrity of the ingredients.

Justin and Deborah live next to the vineyard with their children Evan and Morgane, and an assortment of dogs and horses. Deborah manages the marketing and the Just Inn, and is very involved in presenting the face of Paso Robles to the wine loving public. There are not enough superlatives for Justin. The wines, the dedication, the passion, are tangibly world class. They have earned every bit of praise, and no doubt, there will be more congratulations to come in the future.

JUSTIN VINEYARDS & WINERY
11680 Chimney Rock Road
Paso Robles, CA 93446
Phone: 805-238-6932
Fax: 805-238-7382
Web: www.justinwine.com
Email: info@justinwine.com
Hours: Daily 10 a.m.-5 p.m., and 10 a.m.-6 p.m.
Case production: 45,000

NADEAU FAMILY VINTNERS

Smaller is sometimes better. Robert and Patrice Nadeau purchased eight acres north of Peachy Canyon Road in 1995. They gutted and refurbished an old barn on the property, and planted Zinfandel, Grenache and Petite Sirah cuttings from some well-known vineyards (courtesy of Benito Dusi, Frank Nerelli and Richard Sauret, and some Grenache Noir from Tablas Creek).

Robert's first release was the 1997 vintage, released in 1998. He uses a variety of oak; French, American, Hungarian, about 30-50% percent new. Their pride and joy are their own Zinfandels, "Home Ranch", "Epic", and "Critical Mass", made from their own dry farmed, head pruned vines, fermented in open top fermenters and pressed off on a small, horizontal press. The wine spends one to two years in a combination of new and neutral American oak, and is then racked, filtered but not fined, before final bottling.

They also source red Rhone varietals from the east and west side, for their RVR (Rhone Varietal, Red) series, a 60/40 blend of Syrah and Grenache. They make a 100% Syrah as well, as well as a 100% Grenache bottling, which spends 15 months in French oak, and source some Central Coast Roussanne to blend with Viognier. They produce a 100% Viognier too, stainless steel fermented, aged sur lees, and spending a year in new and one year old French oak. There is a Paso Robles Barbera, blended with some Syrah from Dr. Michael Gill's Vineyard, and a varietal Mourvedre, also blended with a small amount of Dr. Gill's Syrah. They produce a huge, extracted, Late Harvest Zinfandel, and an Alicante Bouchet sourced from the Sierra Foothills, which they blend a little Syrah into. The simply named Mixed Black is comprised of their estate Zinfandel, Petite Sirah and Grenache, displaying round dark red fruit and spicy character.

The Nadeau's are expanding their production facility, but will always specialize in small lot production (generally less than 200 cases) of premium Paso Robles varietals. The current production is about 2,000 cases, and they plan to top out at about 5,000. Straightforward, quality-not-quantity, and delicious are the operating themes at Nadeau Family Vintners.

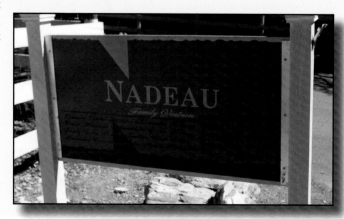

NADEAU FAMILY VINTNERS
3860 Peachy Canyon Road
Paso Robles, CA 93446
Phone: 805-239-3574
Fax: 805-239-2314
Web: www.nadeaufamilyvintners.com
Email: pnadeau@tcsn.net
Hours: Weekends and most holidays Noon-5 p.m.
Case production: 2,000

Art (sadly, now deceased) and Lei Norman settled in Paso Robles in 1971, spurred by Art's fond memories of his father and grandfather making wine at home during his childhood. Art was an engineer working in the aerospace industry when their search for the right site, carried out on weekends, finally landed on the 40-acre Adelaida site that is Norman Vineyards today. They were part time residents until the mid-80's, farming barley, safflower and oat hay along with grapes. They planted to Chardonnay, now since grafted over to Cabernet Sauvignon and Cabernet Franc, but the majority of their estate is planted to Zinfandel, Cabernet Sauvignon, Cabernet Franc, Merlot, and a tiny plot of Barbera right around the house.

For twenty years, they sold this superb fruit to a "who's who" of Paso Robles wineries. In 1992, with Paso Robles emerging as a larger player in the wine industry, Art enlisted Robert Nadeau to help them produce their first vintage of Cabernet Sauvignon, Zinfandel, Barbera, Chardonnay and a proprietary blend called "No Nonsense Red." The wines were quickly and warmly received, and soon they not only were using all of their own fruit, but also needed to source from other growers. These are mostly small, family owned westside vineyards, but they source some Chardonnay from Edna Valley, and old-vine Zinfandel from the Guasti property in Rancho Cucamonga. All of their vineyards, as well as a number of vineyards that they purchase fruit from are ably managed by Willie Silva, and his team, since 1997.

Since 2005, well-known Paso Robles winemaker Steve Felten (formerly of EOS Estate Winery) has been at the helm of the winemaking staff. He says he "doesn't have a formula, and am always looking for something new and interesting to add to the lineup..." His philosophy about oak influence is pragmatic, but demanding... "I only work with the best producers that let me choose the forest, (that the oak comes from) usually French or American, and will let me specify the barrel construction..." He uses anywhere from 60% new oak for wines such as "Crescendo," a red Bordeaux blend that spends 20 months in barrels, to 100% new French oak for the Reserve Cabernet Sauvignon, and just about everything in between for the close to 20 varietals that they produce. There are three classic Paso Robles Zinfandels. The "Monster", which is 100% Zinfandel done in neutral oak uprights, as well as spending some time in new and one year old barrels. There is a "Classic," blended with a little Syrah, again mostly finished in the oak uprights,

Rounding out this varied and interesting portfolio are an "Old Vine" (from the Guasti property) Zinfandel from Cucamonga, far south of Paso Robles, in southern California, actually, an area east of Orange County, that was of great historic importance to the California wine industry, but now sadly being taken over by developers. They make a Zinfandel port from this fruit as well and also obtain Chardonnay from Edna Valley; but over 90% of Norman's production is Paso Robles...there is Merlot, Syrah, Petite Syrah, Barbera, a dry Rosado made from Grenache and Syrah, and Pinot Grigio as well. Steve is also expanding into Tempranillo and Nebbiolo from vineyards in the Templeton Gap area.

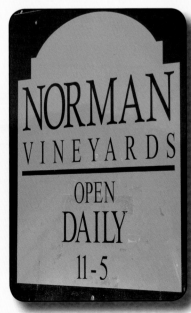

There is a beautiful large tasting room, and the winery, designed with respect for the historical architecture of the area by Mark Jepson, and it seems completely at ease. There is a large outdoor patio surrounded by a brook and intact riparian area, full of oaks, Toyon and mulberry. It is the perfect place to experience Paso Robles and sample from a large palette of beautiful wines.

NORMAN VINEYARDS
7450 Vineyard Drive
Paso Robles, CA 93432
Phone: 805-237-0138
Fax: 805-227-6733
Web: www.normanvineyards.com
Hours: Daily 11 a.m.- 5 p.m.
Case production: 24,000

TABLAS CREEK VINEYARD

The 30+-year association of Robert Haas and the Perrin Family (of Chateau de Beaucastel fame) has always been a productive one. When the Perrins decided to look for a place to establish a new world location for Rhone varietals, it took the Paso Robles story to another level. A California vineyard site had been on the radar screen for years, but in 1987 the search began in earnest, and by 1989 they had settled on a 129-acre parcel in the Las Tablas district of northwest Paso Robles, situated at 1,500 feet elevation. They named it Tablas Creek Vineyard after the small creek running through the property.

In this special place were calcareous, high pH soils, a rugged topography, and near ideal grape growing conditions. The southern Rhone varietals the Perrin family had grown in the Chateauneuf du Pape region of southern France since the early 1900's would thrive here, and Tablas Creek was off and running.

Not so fast, folks...they weren't too impressed with the vine stock available in the USA. That necessitated the importation of vinifera stock from the same genetic source at their vines at Beaucastel. Several clones each of Mourvedre, Grenache Noir, Syrah, Counoise, Roussanne, Viognier, Grenache Blanc, Marsanne and Picpoul Blanc were imported, to duplicate the clonal selection of an established French vineyard. To make a very long story very short, the process of USDA mandated indexing for viruses took three years, but by 1993 they had the first on-site grapevine nursery of any vineyard in California. Planting continues at the estate, with the goal of having 110 acres planted to vines by 2010.

Tablas Creek clones are now available through their commercial nursery partner, NovaVine, and have enriched the vineyards and wines of L'Aventure, Justin, Garretson, Bonny Doon, and long, illustrious list of others. They continue use their own nursery to source material for Tablas Creek. The property was organic-certified in 2003, and is densely planted (1,600-1,800 per acre), severely pruned, and mostly dry farmed.

Winemaker Neil Collins was born in Bristol, England, trained as a chef and moved into winemaking via stints with John Munch at Adelaida Cellars and Ken Volk at Wild Horse. He met Robert Haas and the Perrins at the very beginnings of Tablas Creek, offered his services, and interned for a year at Beaucastel. He has been winemaker and overseer of the organic vineyard operations since 1998. They have recently installed solar panels in the vineyards to supply at least half of the winery's power needs.

Each varietal is harvested at total ripeness, and fermented separately using indigenous yeasts. The whites are gently whole cluster pressed. The Roussanne is partially barrel and stainless steel fermented, the Marsanne, Viognier and Grenache Blanc see no oak. The reds are sorted and destemmed, with juice and whole berries moved to open top fermenters, for twice-daily punchdowns. Post fermentation, they are pressed, blended, and aged for one year in 1200-gallon French oak foudres. Some of the reds see additional barrel aging, but again, the focus is on expression of *terroir* and fruit, not oak.

The list of wines is long; Cotes de Tablas, a Grenache, Syrah and Mourvedre blend, Esprit de Beaucastel, based on Mourvedre, a varietal Mourvedre, Panoplie, another very limited, Mourvedre based cuvee, a varietal Syrah, a Las Tablas Estates Glenrose Vineyard Syrah/Rhone Blend, and a new varietal Tannat, from a recent, groundbreaking planting. The whites include: Cotes de Tablas Blanc, mostly Viognier, Roussanne, and Marsanne with a little Grenache Blanc, varietal Picpoul Blanc, and Esprit du Beaucastel Blanc, Roussanne based, and a varietal Roussanne. They also produce a very highly regarded Mourvedre based rose, and a Vin de Paille ("straw wine") patterned after the classic French dessert style wine, made from grapes dried on straw. With 15,000 cases at present, the winery's total production will probably rise to 22,000 cases by 2010.

Their wines have gotten the attention of the international wine press from the beginning, and continue to be a darling in wine circles. The fact that this winery has been a benchmark influence and direction for Rhone varietals in Paso Robles is irrefutable, and their innovations continue to evolve and flourish.

BEAUCASTEL GRENACHE VINES

Tannic acid fruits from the Vitis species,
Flourishing in gravelly and clay Rhone River alluvium.

Attacked by fungi, insects, and mammals;
Protected by human chemical and physical weapons.

Blended with co-habitating musts and yeasts,
Fermented in epidermis; with yeasts producing alcohol.
Pigments bleed from skins and add flavors;
Aged in French oak for years of peaceful turmoil.

Juices racked, decanted and aged again in damp
Cool rock-walled caves to reach peak quality.

French proprietors join with new world partners;
Import the scion & rootstock vegetation to California for
Propagation and vineyard establishment in comparable
New World terroirs; blind and open tastings bring
Acclaim; a marriage blessed by Bacchus and Ceres.

Photos 23-24: Chateau de Beaucastel, Grenache vines and limy soil north of Chateauneuf du Pape, France (July 1998).

TABLAS CREEK VINEYARD
9339 Adelaida Road
Paso Robles, CA 93446
Phone: 805-237-1231
Phone: 805-237-1314
Web: www.tablascreek.com
Email: info@tablascreek.com
Hours: Daily 10 a.m.- 5 p.m.
Case production: 15,000

VILLICANA WINERY

A thwarted attempt at culinary school was Alex Villicana's entrance into the wine business. The school never opened, but his passion for food and wine led to an early 1990's internship at a Paso Robles winery, which in turn led to UC Davis viticulture and enology classes and trips to the wine regions of Europe with wine critic Bob Balzer. Not a bad start...but he wanted to purchase and eventually grow some wine grapes!

In 1992, he secured a load of leftover Gamay grapes, and took advantage of the offer of three tons of Adelaida Hills Cabernet Sauvignon the very next year. More fruit followed from the same source, and with the help and encouragement, a budding Paso Robles winemaker began to bloom. A marriage ensued, as well, when Alex married Monica in 1995, the same year they purchased 72 acres on the westside of Paso Robles.

After considerable soil, water, rootstock and clone analysis, they personally planted Cabernet Sauvignon, Zinfandel, Merlot, Mourvedre, Syrah, Grenache, Petite Verdot and Cabernet Franc on 10 acres. Recent plantings include Cabernet Franc and Tempranillo. In 1999 they crushed the first vintage of their own estate fruit. They currently produce about 1,000 cases, and will max out production at 5,000. The idea for them is to remain small, and focus on artisanal quality.

Their current release lineup consists of a Syrah based blend (with 20 percent Grenache and Mourvedre) sourced mainly from their own and the Rolph Vineyards on Adelaida Road. Alex used 30 percent new French oak barrels for 19 months, to mellow out this brawny, but complex wine. There is also a dry Rose based on Mourvedre, Grenache and Viognier, from their own, Nevarez vineyard, and westside sources. It is taken off the skins after three days, and sees seven months of French and Eastern European oak. His Winemaker's Cuvee is a Cabernet Sauvignon/Syrah/Merlot blend, all estate fruit. They have veered away from white varietals to concentrate on the red wines that succeed so well in Paso Robles.

The Villicana's have recently sprouted two adorable scions: Alexander (Jr.) and his younger sister, Gabrielle. The tasting room is charming, and you can find Alex pouring his wonderful wines there most weekends. Really a neat story and a great place to visit.

VILLICANA WINERY
2725 Adelaida Road
Paso Robles, CA 93446
Phone 805-239-9456
Fax: 805-239-0115
Web: villicanawinery.com
Email: villicanawinery@earthlink.net
Hours: Thurs.-Sunday 11 a.m.-5 p.m.
Case production: 1,500

TEMPLETON GAP (approximately 35,600 acres): cooler region II climate, and most maritime climate with pronounced marine influence through Santa Lucia Range wind gaps west of Templeton, mountain slopes and old fan and alluvial terrace deposits of the Salinas River, bedrock middle and lower members of Monterey Formation and Quaternary river deposits at lower elevations, largely calcareous soils of both residual (bedrock) and depositional (alluvial) origin, with oak woodlands (Elliott-Fisk, 2007).

BELLA LUNA WINERY

You do not usually come to the USA, let alone the very warm, dry (at least in the growing season) Paso Robles area, and expect to find dry farmed vines. Like Dave King at Vista Del Rey in the Salinas River west area, Bella Luna owners Sherman Smoot and Kevin Healey believe that this type of viticulture results in greater complexity and flavors that are more concentrated. The fascination with Mediterranean varietals is here too, with their first release of Sangiovese in 2001.

Healey has spent his life in the vineyards around Paso Robles, and Jim Smoot says he's the best dry farmer in the country. Kevin is a Vietnam war veteran, and Sherman trained as Navy pilot during the same period. Both grew up in the Paso Robles area, and shared a passion for wine. After his stint in Vietnam, Kevin came home to roost in the Adelaida Hills, receiving tutelage in dry farming from the local master, Mel Casteel, and spent a decade learning his craft as an organic viticulturist. This was the beginning of his career as a winemaker, as well, using, well, just about anything that would ferment.

Onward in this journey he worked with the then family-owned Pesenti Vineyard and Winery. Working there with head-trained mature Zinfandel vines with the founder Frank Pesenti, he also developed a friendship with Pesenti's winemaker, Frank Nerelli. Working closely together, they produced some of the most sought after wines in California, all from dry farmed, head pruned vines. Kevin worked here as the vineyard manager for 19 years until the winery was sold to Turley in 2000.

Sherman had returned to the US after the Vietnam War to become a commercial aviator. Expanding his palate during his global travels, and utilizing his degree in biological sciences from Washington State University, he began to avidly explore the fields of enology and viticulture. He would bounce his growing base of ideas and new techniques of the sometimes skeptical Healey, but the foundation for the next phase of his life was firmly laid down. Due in part to the influence of his old world Italian stepfather, Dominic Marietta, who made his own wine as well, Sherman's passion for Italian varietals, particularly Sangiovese, shines bright. Kevin is in charge with bringing all the best of this temperamental vine out... Sherm says, "he is an absolute genius in the vineyard..."

The five-acre Bella Luna estate has several different soil types. These are the only dry-farmed, head-trained Sangiovese vineyards in California. The Rockpile, composed of alluvial river bottom, gives the most intense fruit. Their clay loam soils are well suited for dry farming, and where they get the highest yielding crop. They grow their Cabernet Sauvignon from the sandy loam section on the southeast hill, as well as the loam soil on the bottomlands. All vineyards are farmed organically, and almost all are head pruned, except for one-half of the Cabernet Sauvignon vines, which are vertical cordon trained without trellising. The total vine count is still under 2,000, planted at medium density, 12- by 7-foot spacing. They benefit greatly from the classic Paso Robles diurnal temperature fluctuations, producing wines that are rich, balanced, concentrated and evocative. Their flagship wine, the Estate Riserva, is about equal proportions estate dry farmed Sangiovese and Cabernet Sauvignon, produced in tiny quantities in numbered bottles. Sherman's Navy F-4 pilot days are honored with their other wine, Fighter Pilot Red, a blend of their Sangiovese and some dry farmed

Zinfandel sourced from other nearby vineyards. They will limit their production to about 500 cases at Bella Luna. These are rare wines worth trying.

BELLA LUNA WINERY
1850 Templeton Road
Templeton, CA 93465
Phone: 805-434-5477
Fax: 805-434-5479
Web: www.bellalunawine.com
Hours: By appointment and during special events
Case Production: 400

CASTORO CELLARS

Niels and Bimmer Udsen rather backed into the wine business. Niels started making wine, selling it locally, and gradually acquiring equipment and property...to some remarkable results. Originally learning to make wine with his father-in-law in Denmark (he has known Bimmer since he was 8 years old!), he really caught the bug (and the winery name) during a visit to Italy after high school.

The son of a Danish immigrant, Niels started at Estrella River Winery, under the tutelage of Tom Meyers, learning it all from the ground up. He made wine in his spare time, his wife Bimmer delivered the finished product and kept the financial books...but little by little, the brand (named after his Italian nickname of "Il Castoro"- the beaver) began to grow.

Launching Castoro Cellars, in 1983, they owned no property or buildings, and were buying grapes from dozens of different vineyards, renting space to make the wine wherever they could find it. Their philosophy of high quality, approachable and affordable wines for all was taking root. The variety of vineyard sources enabled them to establish a high profile local brand that resonated with loyal customers.

Niels also started up a custom crush business, buying a press and a mobile bottling line for small wineries to lease out...this would also provide the financial base for them to purchase new and better equipment, and therefore make better wine, than sales of the brand alone would allow. They bought a winery site, in San Miguel, purchased vineyards and planted their own vines. Castoro Cellars tasting room is a very busy welcoming place on the Paso westside, very family friendly and "green."

Their own vineyards are increasing yearly, with the final intention of making single vineyard wines. They also source grapes from 20 vineyards, and Tom Meyers is still in the picture, directing winemaking here since 1990, along with assistant winemaker Mikel Olsen. Mesa Vineyard Management oversees all the vineyards, and continues to develop the over two-decade-old relationships with a prime palette of growers in the Paso Robles AVA.

They have some interesting own-rooted vines in a number of areas: on the east side in Dos Vinas, and the Blind Faith vineyard (bought by Nils without Bimmer's input during a visit to Denmark) and the westside Cobble Creek home ranch old vine Zinfandel; head pruned, dry farmed, and consistently produces some of the finest Zinfandel in Paso Robles. They planted several different Zinfandel clones, some Primitivo, and Carignane on Bethel Road, all organically farmed as well.

These are delicious, rich, fat wines. They produce the full range of core varietals, Chardonnay, Sauvignon Blanc, Cabernet Sauvignon, and Merlot, but it gets more interesting...they also produce an estate bottled Viognier, several stunning single vineyard Zinfandels, a reserve style Bordeaux blend named Due Mila Quattro, a "mystery" red blend (it changes every year), Ventuno Anni, a Petite Sirah, and some amazing late harvest and fortified wines from Zinfandel and Muscat Canelli. They also sell their own grape juice, a great alternative to sugary sodas, for kids and parents alike...I wish more wineries did this.

Castoro Cellars has become a lynch pin winery that espouses some wonderful ideas about the wine lifestyle: approachability, affordability and sustainable environmental practices. This is a recipe for long-term enjoyment and success!

CASTORO CELLARS
1315 N. Bethel Road
Templeton, CA. 93465
Phone: 805-238-0725
Fax: 805-238-2602
Web: www.castorocellars.com
Email: info@castorocellars.com
Hours: Daily 10 a.m.-5:30 p.m.
Production: 30,000 cases

DOCE ROBLES

Jim and Maribeth Jacobsen are natives to the world of viticulture. They are third generation family farmers from the Fresno area, where their families grew wine grapes, as well as almonds and peaches. Wine tasting lead them to Paso Robles to start with, and by 1996 had planted 40 acres to red varietals. The day-to-day running of their winery is very much a family affair, down to Maribeth's mother, artist Gail Hansen who does the beautiful paintings for the labels. Jim is the winemaker...and his goal is to grow and make wonderful wines from Paso. Almost all of the production is estate bottled, except for a Nebbiolo, Sauvignon Blanc and a Viognier.

Doce Robles is Spanish for Twelve Oaks, derived from a scene that featured a dozen oaks on an etched glass window at the house on the property. This is a classic, pastoral Paso Robles scenario: rolling hills dotted with oak, friendly dogs (Duchess, Syrah and Ellie) lazy cats in a basket on the tasting bar (Grayson) and Trickem the horse. Quite the menagerie, but it's all good; not to worry, Trickem is outside to welcome you.

Winemaking here is simple, noninterventionist, and to the point. Jim says that "most of the real work is done in the vineyard" and you can't make a silk purse out of a sow's ear...They utilize organic practices, and use food-grade fertilizers when necessary to correct nutrient deficiencies in the vineyard. All the wines are totally handcrafted in small open top fermenters, gravity fed, pumped only twice from field to bottle. They spend on the average two years in French and American oak. They bottle estate-grown Barbera, Zinfandel, Cabernet Sauvignon, and Syrah, which is fast becoming their

flagship wine. There are some great red blends, too: Robles Rojos (a Bordeaux blend that sells out quickly), a Sy-rific Red...Cabernet and Syrah here, a late harvest Zinfandel, a "normale" (Chianti term for regular bottling) Zinfandel, and a delicious Rose Robles from Syrah, Grenache and Cabernet Sauvignon. Sauvignon Blanc, Viognier, Merlot, Petite Sirah, and Nebbiolo round out the rather large portfolio of bottlings. Jim just can't sit still, we figure.

It says on their web site that they are known as the party winery, which they are totally happy about. All their many tasting room visitors are too!

DOCE ROBLES
2023 Twelve Oaks Drive
Paso Robles, CA 93446
Phone: 805-227-4766
Fax: 805-227-6521
Web: www.docerobles.com
Email: docerobleswinery@tcsn.net
Hours: Daily 10 a.m.-5:30 p.m.
Case production: 1,500

Fratelli Perata

Italian heritage runs right through the heart of Paso Robles, and Fratelli Perata (Italian for Perata Brothers) is a great example. This small, family owned winery is an outgrowth of several generations of winemaking and agricultural roots, going all the way back to Genoa, Italy. The heirs to this tradition, Gene and Joe Perata, now farm their 25-acre estate and make about 2,000 cases of ageable, very *terroir* driven wines.

Let's go back to 1930, when Joe and Gino's father, Giuseppe Antonio, came to the U.S. from Genoa, at the age of 16. Immediately immersed in the agricultural roots of Ventura County, he soon felt the pull of generations of winemakers in his blood, and started making his own wine. His sons, Gino and Joe, haven't fallen far from the tree, so to speak. Gene helped out as a youngster during the vintage, pressed on to classes at UC Davis, and, after meeting and marrying Carol in Washington, began raising a family.

Beginning in 1972, they began a long, meticulous search, from Washington state to Temecula and finally to Paso Robles. Besides the climate and calcareous soils, there was an immediate resemblance to their ancestral homeland. There were also the rolling hillsides, ideal for viticulture. In 1977, they purchased a barley ranch from Leo Shetler, one mile west of the Highway 101.

They proceeded to personally survey and clear this 40-acre parcel, and began planting Zinfandel in 1980. They planted 520 vines to the acre, 25 acres at a time, one bare root vine at a time, two feet deep, with a one-person auger. Family members all helped, first with this chore, and then with the installation of the drip irrigation, trellising and fencing. All of this before the days of ATV's...they call it "putting your shoulder into it" when referring to hard work, but it is an accurate description of the kind of backbreaking labor that planting a vineyard entails.

They sold fruit to Martin Brothers and Twin Hills. Although there was a grape glut at the time, they were encouraged by the response of the winemakers to the big, gutsy Zinfandel wines that their fruit produced, and they became a bonded winery in 1987. The family built the winery, and they released their first wine, a few hundred cases of Merlot. They also marketed it themselves, and soon were making and selling more wine, and selling less fruit. The awards were coming in; Paso Robles was beginning to get a lot of attention.

They continue to make wines in a very traditional, hand made style. Ultra-low, two ton per acre yields, mostly dry farmed once the vines had reached maturity, hand punch downs, extended aging in new French and American oak barrels, unfined and unfiltered, hand bottled and labeled. There is an Estate Nebbiolo that spends 3-4 years in barrel, an Estate and a Reserve Cabernet Sauvignon, made from vines planted in the 1970's, that also spend a great deal of time in French oak. There is a Syrah, a Petite Verdot, Merlot, a fantastic Sangiovese, and, of course Zinfandel.

This is a tremendous family story stretching back generations, and sure to continue as a fine example of classic Paso Robles style and quality. Don't just wander in, they need to know you are coming, but it is worth the extra trouble.

Fratelli Perata
1595 Arbor Road
Paso Robles, CA 93446
Phone and Fax: 805-238-2809
Web: www.fratelliperata.com
Hours: By appointment only
Case production: 2,000

Peachy Canyon Winery

The Beckett family has blossomed in the last 25 years, and Peachy Canyon has become a national favorite of consumers and wine critics alike. From a very modest beginning with a bit of Benito Dusi's Zinfandel, produced at a winery adjacent to the Beckett's family home, to a 46,000 case, globally recognized leader in ultra-premium red varietals from Paso Robles. They own about 100 acres spread around four main vineyards: Mustang Springs, Mustard Creek and the Snow vineyard in the Adelaida Hills, and the Old School House vineyard in Templeton Gap.

There is a tangible passion in Doug Beckett's voice when he speaks about his life's mission in Paso Robles. He and his wife Nancy (who, in addition to her role at the winery, is a well-known dance teacher with over 300 students) were schoolteachers until the early 1980's. Their friendship with a Paso Robles walnut grower grew into their eventual move out of Los Angeles with their children, to make their first hundred cases of wine at the Tobias winery next door. This partnership ended in 1988, and while continuing to make wine, the Beckett's purchased their first vineyards in 1997.

Winemaker, and son, Josh Beckett grew up with in the land that his fruit comes from, and seems to have an innate sense of how to handle each different lot of fruit. Josh, who is a self-taught winemaker, was mentored by his dad, Tom Westberg, Florence Sarazin and Robert Nadeau. He has an artisan's viewpoint about winemaking, "we treat a 50,000 case run like a 200 case run..." using just enough oak (French, American, and eastern European) to complement the wine but not overcome it. They estate bottle Zinfandel, Cabernet Sauvignon, Merlot, Sangiovese, Para Siempre ("for always") alluding to the eternal marriage of the classic Bordeaux varietals, Petite Sirah, Cabernet Franc, Syrah, and Zinfandel Port. The three estate bottled vineyard designates lead off their Zinfandel portfolio; Snow, from a warmer, earlier ripening vineyard site, Old School House, from the vineyard that surrounds their tasting room, and Mustang Springs, which is blended with a little Petite Sirah, from their vineyard planted to old and new vines. There is a Mr. Wilson Zinfandel from the Wilson's Twin Oaks Vineyard, Incredible Reds numbered bin series, and Especial, a blend of their own estate Zinfandels.

Josh makes Vesuvio, a Sangiovese/Cabernet Sauvignon/ Merlot Blend, and Jester, a Bordeaux/Rhone blend...most of the wines are briefly cold soaked, different lots fermented separately, and gently pressed off, mostly in their three ton basket press. He makes a couple of Port styled wines, from the Macgregor vineyard, both are tawny in style and comprised of several different vintages. Some newer projects include a co-fermented Syrah/Viognier blend, and a Grenache from Katherine's vineyard off Creston Road on the east side of the Salinas River.

The tasting room is now located at a historic 1886 Old Bethel School House on Highway 46 west, and there is an aptly named "Peachy Too" tasting room on Nacimiento Lake Drive. This is very much still a family run operation, with the various branches running the sales and marketing machinery, managing events, running two tasting rooms, all this in addition to making state of the art Paso Robles wines. Doug says..."the world is full of dreamers, but sometimes they come here to live them..." We couldn't have said it better.

PEACHY CANYON WINERY
1480 N. Bethel Drive
Templeton, CA 93465
Phone: 805-237-1577
Fax: 805-237-2248
Web: www.peachycanyon.com
Email: info@peachycanyon.com
Hours: Daily 11 a.m.-5 p.m.
Case production: 46,000

SUMMERWOOD WINERY

This stunning winery and "four diamond" nine-room inn opened in 2001, after being purchased by the Fukae family of Osaka, Japan. They rebuilt and refurbished the 35-acre property, planted with Cabernet Sauvignon and Syrah.

The state of the art winery has just hired the master Australian winemaker, Chris Cameron. They utilize the most gentle methods for grape handling: gravity feeds, basket presses, no mechanical pumping of wines or must...production is capped at 5,000 cases, and focuses on westside Paso fruit. They sell all of their small lot production over the web or from the tasting room. "...that's what makes it attractive for people...," says Tom Colero, the tasting room manager.

They produce a Denner Vineyard Syrah and a James Berry Vineyard Syrah, a varietal Viognier, and a 100% Roussanne. Their flagship red and white Rhone blends are called Diosa, a term for goddess. Diosa Blanc is Viognier and Roussanne, the red Diosa is Grenache, Syrah and Mourvedre. The wines are all aged in large oak puncheons in their newly remodeled barrel room.

Cameron staying away from the fruit-obscuring toast and vanilla aromatics that too much new oak can bring. They also produce a Merlot, a red Bordeaux blend called Sentio, and a private reserve Cabernet Sauvignon. There is talk of a Chardonnay to be produced soon...and there is a red Port style wine, as well, that is in high demand. The wines are not inexpensive, but nothing good ever is. The Fukae family, and the team at Summerwood have done a marvelous job of re-inventing this unique property, and Chris Cameron's wine releases promise to receive the full attention that they deserve.

SUMMERWOOD WINERY
2175 Arbor Road
Paso Robles, CA 93446
Phone: 805-227-2541
Fax: 805-227-3516
Web: www.summerwoodwine.com
Email: info@summerwoodwine.com
Hours: March-October 10 a.m.-6 p.m.
        November-February 10 a.m.-5:30 p.m.
Case production: 5,000

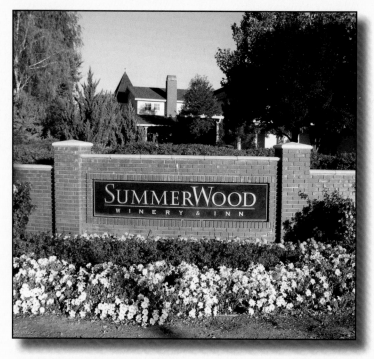

TREANA WINERY

The *terroir* of the Templeton Gap in Paso Robles inspired the name...recalling the trinity of natural elements; the sun, the soil and the ocean, that make this part of the grape growing world so unique. The Hope family began planting vineyards in the area in 1978. Over two decades of experience has taught them not only where, but also how to grow superb quality grapes.

Beginning with 1990 vintage, the Hope ranches became the sole source of fruit for the well-known Liberty School Cabernet Sauvignon. In 1996, they founded Treana Winery, cultivating some of the best examples of red grapes for Paso Robles: Syrah, Cabernet Sauvignon, Petite Sirah, Grenache, Petite Verdot, Mourvedre and Merlot. They source grapes for their white wine, from the Mer Soleil Vineyard in the Santa Lucia Highlands, in the Monterey Region north of Paso Robles.

Founding winemaker Chris Phelps and the Hope's son, Austin, the current winemaker, began by pulling out old vineyards and developing the new ones using the newest techniques; much higher density of vines in particular. They also use clones of Syrah from Hermitage and Chateau de Beaucastel (see the profile on Tablas Creek about this...) as well as Cabernet clones from the St. Emilion region of Bordeaux.

Austin Hope began working in his family's vineyard when he was just eight years old, going on to graduate from Cal Poly, San Luis Obispo with a degree in viticulture and crop science. He is head of viticulture and winemaking for both Treana and Liberty School wines, and studied under Chuck Wagner of Caymus...assisting Chris Phelps during his tenure at Dominus. Quite a pedigree for such a young face!

The Treana White is a blend of Marsanne and Viognier, the majority fermented in neutral French oak, with 10% stainless steel added in to accentuate the high-toned floral characteristics, and spends eight months in barrel. Treana Red, a Bordeaux/Rhone hybrid, is based on Cabernet Sauvignon, with the additions of Syrah and Merlot. Individual lots are fermented separately and then barreled down in 100% French oak, of which 60 percent are new. After blending, it spends over a year in barrel, with minimal racking and filtration just prior to bottling. The continuously outstanding quality of the Treana Red and White wines have been keenly instrumental in placing Paso Robles (and the Central Coast in general) wines on the best wine lists in the United States and globally.

The Liberty School wines have an equally fine reputation as some of the finest entry-level varietal wines in the genre. They source fruit for the Chardonnay primarily from the Monterey/Santa Lucia Highlands, and is given a very similar treatment (limited new French and American oak, more stainless steel) as the Treana White...they blend in a dollop of Viognier at times, too. The Cabernet and Syrah are sourced from small family owned Paso Robles vineyards, given top flight winemaking (small lot fermentation in open tops fermenters, medium lots go into pump over fermenters, manual punch downs, and aging in predominantly new-to-four year old French oak barrels. The Cabernet sees a little new oak as well...they minimally rack and filter the wines prior to bottling.

There is a new chapter, too... Austin Hope Winery, which specializes in proprietary Rhone style blends sourced from Paso westside vineyards. Westside Red is a Syrah/Mourvedre/Grenache blend, with individual lots fermented and barreled down separately. It sees about 35% new, 65% one-to-two year old French oak, and is designed for current consumption...the Westside White, sourced from

Monterey, is a Roussanne, Viognier and Grenache Blanc blend of stainless and French oak aged lots, also made with early, easy, but interesting, consumption in mind.

The flagship of the new label is the Hope Family Vineyard Syrah, planted with four different clones of Syrah at very high density. It is made using small lot open top fermenters, and 16 to 18 months in new French oak barrels, with minimal racking and filtration. Austin makes his own Roussanne as well, again from the Mer Soleil vineyard in Santa Lucia Highlands, 100% barrel fermented in new French oak, with sur lies aging for 9-10 months.

Treana, Austin Hope, Liberty School...a triad of success for the Hope family and a gateway into outstanding wines of Paso Robles.

TREANA WINERY
P.O. Box 3260
Paso Robles, CA 93447
Phone: 805-238-6979
Fax: 805-238-4063
Web: www.treana.com
Email: info@treana.com
Hours: by appointment only

WILD HORSE WINERY & VINEYARDS

This groundbreaking winery was founded in 1981 by the energetic Ken Volk, who came to Templeton looking for groundwater, low vigor soils and a pastoral atmosphere. Ken built the winery and the brand into one of the flagship wineries of the Central Coast, with many of his Paso Robles AVA wines, Cabernet Sauvignon, Merlot, Zinfandel and Syrah. He also made a name for himself sourcing fruit from throughout the Central Coast appellation for what would become one of the coastal areas best known Pinot Noir

Photo 25: Geneva double curtain trellis system used with Pinot Noir grapes at Wild Horse Vineyards (July 2000).

and Chardonnay. Ken sold the winery in 2003 to Peak Wines International, and well-known Aussie winemaker Darryl Groom's talented team, but his eclectic perspective on wines and vines remains. Besides their nationally distributed core varietals, there is the "Unbridled" line of vineyard designated wines from Pinot Noir, Chardonnay and Syrah, and, "Cheval Savage," (French for Wild Horse), their highly sought after reserve Pinot Noir. In addition to all of this (well received and critically acclaimed) commotion, they also produce no less than 22 other varietals, spanning the globe, but with a distinctly Italianate/Mediterranean perspective; Malvasia Bianca, Pinot Grigio, Arneis, Dolcetto, Sangiovese, Tocai Friulano, Primitivo, Lagrein, Tempranillo, Trousseau, Valdigue, Verdelho, Vermentino, Blaufrankisch, and nearly a dozen other wines.

The vast diversity of climates and soils in the Paso Robles and Central Coast AVA's allow them to source and crush close to 30 different varietals each harvest. If it is growing in California, chances are Wild Horse is working with it. This wide range of experimentation leads one not only on a unique tasting experience at the winery, but engenders and fosters a truly multi-faceted, cross-trained team. They learn more about making Pinot Noir by experimenting with Lagrein, an obscure red varietal originally from the Trentino-Alto Adige region of northeastern Italy, and the Chardonnay improves by experimenting with Trousseau Gris, a mutation of the Trousseau grape from the Jura region in France. Not your everyday skill set building exercise, but one that seems to be working.

They have put Paso Robles Merlot on the map, and the recent vintages exhibit a typical spice rack approach to blending, with the addition of splashes of Dolcetto, Blaufrankisch, Malbec, Cabernets Franc and Sauvignon. Their Cabernet Sauvignon is sourced from 18 different vineyards, and is mostly Cabernet Sauvignon with a "skosh" of Malbec and Merlot. Their Zinfandel is not in the much-ballyhooed "Gorilla Zin" style of many a Paso Robles winery, but reflects a European influence, almost Claret-like, but with some wonderful briary old vine character from the Enz Vineyard near Hollister, which was planted in the 1880's. They have released a delicious Syrah, sourced from three vineyards, and is blended with Mourvedre.

All the Wild Horse fruit is hand harvested, and Groom's team has followed Ken Volk's philosophy of moderate oak influence...lots of different sources here, too: French, American, Hungarian...use of cold soak for the red wines, open and closed top fermenters, rack and return, all lots kept separate until blending. These are hand made wines with extreme care at all stages. This winery has gone onto its next phase, and faces the 21st century galloping on all four legs.

WILD HORSE WINERY & VINEYARDS
1437 Wild Horse Winery Court
Templeton, CA 93465
Phone: 805-434-2541
Fax: 805-434-3516
Web: www.wildhorsewinery.com
Email: info@wildhorsewinery.com
Hours: Daily 11 a.m.-5 p.m.
Case production: 138,000

WINDWARD VINEYARD

Pinot Noir is Marc Goldberg and Maggie D'Ambrosia's passion. The are seeking "….to replicate a fine French Burgundy…" Marc unselfconsciously states, and many would say, that they have come a lot closer than many to this vinous Holy Grail. Windward is dedicated to Paso Robles Pinot Noir, period. "...I allow the vineyard to make the wine for me..." and one wine, from Pinot Noir, is all Marc Goldberg makes at this very cool, unique site in the Templeton Gap. At Windward, *terroir*, a word much bandied about these days, is taken very seriously. Taken literally, this French concept is all about the convergence of soil, climate, aspect and varietal.

Marc and Maggie purchased their fifteen acre vineyard in 1989, based on the *terroir* and the mesoclimate of the west side Templeton Gap. The cooling Pacific breezes, calcareous soils and significant 40-50 degree diurnal temperature range were perfect. From the inception the idea was total control in the "monopole" concept of fine French Burgundy. They planted it to some seriously pedigreed clones of Pinot Noir; cuttings from Hoffman Mountain Ranch, which were "suitcase" clones taken from the Burgundy Grand Cru Clos de Beze, Sanford and Benedict's Martin Ray clones, which came from Romanee Conti, Clone 4 Pommard and Clone 13 Bien Nacido. In this day of mass plantings of similar, genetically identical clones, this approach means "…you can taste the layers of complexity…" according to Marc. Their first release in 1993 sold out immediately, a trend that continues today, fifteen vintages later.

"Our wines are made in a traditional style...no fining, no filtration, no acid, no manipulation..." The fruit is cold soaked for about five days, and manually punched down in stainless steel three times a day during fermentation. The free run wine is racked off to another tank and the cap is gently pressed. The two lots are blended together prior to malolactic fermentation, and off it goes to 100%

BACCHUS' LYRE

Oh, what beautiful music
You play, sweet Bacchus!

Sweet juices made from the fruits of
Ancestral genetic flavor organisms.

Store chlorophyll in the leaves to manufacture
Sugars that translocate to the reproductive fruit.

Add fermenting yeasts to transform the
Carbohydrates into mind-altering alcohol.

Serve the god's silken nectar in crystal decanters;
Inhibitions vanish as Bacchus strums his mythical lyre.

Google Earth image of Doce Robles Winery, Summerwood Winery, and Windward Vineyard; located north of Highway 46 West.

French oak, custom Seguin Moreau Grand Cru barrels, one third new, one third one-year old, and one third two-year old. The production is limited to 2,000 cases and the wines sell out every vintage. They are definitely walking the talk about Pinot Noir.

These are unique wines from a singular mesoclimate; in many ways a departure from the Paso Robles norm. Windward's wines aging ability are a local legend; the wines mature beautifully into their second decade. We wish them continued worldwide recognition and the success they deserve amid their Rhone and Bordeaux styled neighbors, and look forward to drinking these marvelous Pinot Noirs for years to come.

WINDWARD VINEYARD
1380 Live Oak Road
Paso Robles, CA 93446
Phone: 805-239-2565
Fax: 805-239-4005
Web: www.windwardvineyard.com
Hours: Daily 10:30 a.m.-5 p.m.
Case production: 2,000

ZENAIDA CELLARS

Eric Ogorsolka's family purchased their 95 acre westside property, Rancho Encino, in 1988, after relocating here from the northwest USA. At first they modestly planted five acres, but quickly realizing the quality of the fruit, their vineyards have grown to 65 acres, planted to Zinfandel, Syrah, Petite Sirah, Viognier, Cabernet Franc, Cabernet Sauvignon, Merlot, Pinot Noir and Chardonnay. The Zenaida label was launched in 1993. They have a low impact philosophy; the winery now occupies the original 100 year old homestead on the property. Their rustic, early California themed winery fits in perfectly, overlooking ancient oaks and the vineyard.

Eric's enological leanings began as a hobby...with a BA in biology from Cal Poly SLO, he experimented with fishery science as a career, but decided to pursue the kinder, gentler winery/ranch owner lifestyle. His wines have a European balance and softness to them, he uses a little intervention as possible, gravity flow from vineyard to bottle, some cold soaking, manual punch downs in open top fermenters, not a lot of new, but mostly French oak. His "ZC Red" is a Bordeaux blend, based on Cabernet. "ZC White" is based on Chardonnay and Viognier...they produce a Cote Rotie inspired red blend called "Zephyr Estate Cuvee." An unusual blend of (separately fermented) Zinfandel, Syrah and Viognier, and estate Syrah, Cabernet Sauvignon, and "Fire Sign," a Cabernet-based Sauvignon/Syrah/Zinfandel blend, that sees new French oak for 30 months, and a year of bottle age before release. Not your usual Paso Robles lineup by any means. They produce 3,000 cases now, plan to expand to 10,000 cases over the next several years.

There is a publicly available, stunning 1,500 square foot "loft" suite that overlooks the winery...a trip to Zenaida, named after a Latin term for "mourning dove" is a glimpse into special Paso Robles *terroir* vision and an enchanting journey to fine wines.

ZENAIDA CELLARS
1550 Highway 46 West
Paso Robles, CA 93446
Phone: 805-227-0382
Fax: 805-227-8349
Web: www.zenaidacellars.com
Hours: Daily 11 a.m.-5 p.m.
Case production: 3,000

"QUICKLY, BRING ME A BEAKER OF WINE,
SO THAT I MAY WET MY MIND AND SAY SOMETHING CLEVER."
ARISTOPHANES

PASO ROBLES WILLOW CREEK DISTRICT (approximately 21,300 acres): a mountainous area within Templeton Gap, cooler region II climate and along with Templeton Gap the most maritime climate with pronounced marine influence through Santa Lucia Range wind gaps west of Templeton, cold air drainage downslope, mountain slopes of Santa Lucia Range (with small area of older terraces), bedrock middle and lower members of Monterey Formation, shallow, calcareous soils of residual (bedrock) origin, mixed woodlands at these higher elevations (Elliott-Fisk, 2007).

CALCAREOUS VINEYARD

From Sioux City, Iowa to Paso Robles is a major leap by any measure. For Lloyd Messer (sadly, now deceased), and his daughter Dana Brown, it meant selling their wine distribution businesses and setting off to forge a dream...of making wine from premium vineyards. Erika Messer, Dana's younger sister, joined the team as well, as hospitality manager. Lloyd and Dana purchased a 442 acre westside estate that possessed the unique (surprise...calcareous... mudstone, shale and limestone-derived) soil profiles they had sought, hired Justin Kahler as winemaker, and they were off and running. While they still have long-term leases with premium vineyards, the goal here is to produce estate-bottled varietals.

Again, here at their vineyard sites, one sees the classic New World scenario of a ragbag of varietals growing with equal success next to each other. The proximity of calcareous sedimentary and volcanic-derived soils to each other gives the winemakers amazing opportunities to source and make a variety of wines. Justin uses native/indigenous yeasts and a barrel-aging program of French and American oak. The varietal lineup includes the ubiquitous Chardonnay, but segues more interestingly into Rhone Whites such as Viognier and Roussanne...their selections of reds includes the prerequisite Zinfandel, but also Syrah and a York Mountain Cabernet Sauvignon. The star of the group right now is the Lone Madrone Pinot Noir, which sells out every year, and stands as a testament to

the capability of mesoclimates within the Paso Robles region to produce world class Pinot Noir. The ultimate goal is to produce luxury Rhone and Bordeaux blends, as well as Burgundian varietals. Quite a tall order, but in typical Paso Robles fashion, they are happily motoring along, selling out their releases every year, and there will be a very bright and successful future ahead for the "twisted calcareous sisters."

CALCAREOUS VINEYARD
3430 Peachy Canyon Road
Paso Robles, CA 93446
Phone: 805-239-0289
Fax: 805-239-0916
Web: www.calcareous.com
Email: service@calcareous.com
Hours: Thursday-Monday 11 a.m.-5 p.m.;
Tuesday and Wednesday by appt.
Case production: 5,000

CASA DE CABALLOS VINEYARDS

Paso is brim full of very, very serious wineries and winemakers. However, there is a twist to the whole picture. To put it succinctly, people come here to fashion and live out their dreams. Sometimes they coin a couple of dreams at the same time. Tom Morgan (a physician) started out making fruit and berry wines...and has obviously come quite a ways from there. He looked all over California for land, and settled on Templeton in 1975. He and his wife, Sheila, came from Orange County after their marriage in 1981, to a one acre plot on Templeton Road, that boasted some walnuts and a couple of Arabian horses.

He established a medical practice, and they immediately planted vines: Pinot Noir, Riesling, then Muscat Canelli, Merlot and Cabernet Sauvignon. In typical independent Paso Robles fashion, they wanted to produce wines that they liked to drink, not too concerned with the completely commercial aspect. It soon became clear, however, that they and their friends would end up in rehab trying to consume all the wine they produced, from what had grown to be a six-acre operation. Tom Morgan's dream of breeding Arabian horses came to fruition in 1984, with the birth of their first foal. The winery was bonded in 1995, and Morgan Farms became Casa de Caballos (House of Horses). The Morgan Farm Arabians are now nationally known, award-winning stock, and the seven wines that they produce are named after individual horses, sporting pictures of these magnificent animals on each label.

Their son, Scott Morgan has entered the fray, assuming responsibilities for sales and marketing. He also works in the vineyard and the winery...and produces his own luxury label as well, the tiny production Scott Aaron wines. They are cutting edge, and gorgeously packaged. The red is a Bordeaux varietal blend based on Cabernet Franc, a Viognier. He is winning awards right out of the gate, too.

All of their wines are hand made. Tom is involved at every level, from "dirt to bottle"...no machine picking, or pruning...fermented in small lots in French and American oak, hand punched down, truly artisanal. They estate-bottle everything, and do some interesting blends; a Pinot Noir with 10 percent Merlot, called Maggie Mae El Nino Red, a 100% Pinot Noir from the first vines they planted, called Periwinkle, that spends 11 months in barrel. There is a Cabernet Sauvignon/Merlot blend: Forgetmenot; and two 100% varietal, single vineyard wines: Ultra Violet Merlot, Chocolate Lily Cabernet Sauvignon. The selection is rounded out by two fruity blends, Lilac Thyme, a Riesling/Muscat/Pinot Noir blush, and Fantasy Riesling, which thrives, unusual for this cool climate grape, at the close to 1300 foot elevation, blended with 25% Muscat Canelli. They sell almost every drop of their total 500 case production out of the tasting room and online...a charming and fitting end to this lovely story.

CASA DE CABALLOS VINEYARDS
2225 Raymond Avenue
Templeton, CA 93465
Phone: 805-434-1687
Fax: 805-434-1560
Web: www.casadecaballos.com
Email: info@casadecaballos.com
Hours: Daily 11 a.m.-5 p.m.
Case production: 500

DARK STAR CELLARS

Norm Benson's web site newsletter rocks. He rants, raves, and brings you into the inner workings of a winemaker's life in an almost scary way. But his wines aren't scary, they're great. He's got the pedigree, too...His father was in the beverage business before it was the wine business, and Dark Star was bonded in 1994; coincidentally, on his father's birthday, December 15.

In 1990, Norm's parents planted a vineyard in Paso Robles, and soon Norm was making the round trip to Paso Robles from his home in Thousand Oaks, becoming enmeshed in the local wine culture. This was a turning point for him, and in 1993, he and some very loyal associates crushed his first thousand pounds of Cabernet Sauvignon from this vineyard in his garage. That turned to five tons of Merlot at a friend's winery, which in turn morphed into 15 tons in 1995, at which point they purchased the current winery property on Anderson Road. His parents sold the original vineyard and moved back to southern California before Dark Star was founded.

Norm's first career was a 25-year stint as a transportation coordinator in the motion picture industry. For the first five years of winery operation, he ran a grueling schedule from the winery to studio and back...In 2000, with the support of his wife and family, he "retired" to devote his full attention to the winery. In the eight years hence, production has gone from 300 cases of Merlot, to 4,000 cases of ultra-premium, award winning wines.

Norm espouses minimal intervention with his wines, utilizing a concept he calls "synthetic gravity" which uses low-pressure pneumatic pumps and nitrogen racking. All the fruit is hand harvested and fermented at low temperatures. The winery's flagship red, Ricordati ("always remember"...a homage to his father) is a classic Bordeaux-style blend that sports a laundry list of awards and Gold medals for the 10 years it has been produced. They produce a Cabernet Sauvignon, a Merlot, a superb Paso Zinfandel, and Zinfandel Port, and an interesting, delicious "Anderson Road" Cabernet/Syrah/ Petite Sirah blend.

At harvest time, they remove 100 gallons of juice from each lot of grapes that comes into the winery and slowly ferment it at 40 degrees (Fahrenheit, of course), adding juice each time they harvest a vineyard. At the end of the harvest, they warm the tank, finish the wine and celebrate with the no doubt very rich, almost cru- Beaujolais style wine, fermented over the previous two months. It is quite appropriately called Celebration.

"Angeli d'Altri Tempi" calls out from the triangle at the top of the Dark Star label. Translated as "angels from other times," this saying symbolizes the collective influence of all the treasured people in our lives. The three panels that encase the Dark Star represent the past, present and future, encouraging us to never forget the impact that these people have had in our lives, the support of the myriad people in our present, and the mystery of who we will meet in the future.

This is a true family operation...it is Norm Benson, his wife Susan, son Brian and daughter Nicole doing it all; farming, viticulture, winemaking, finishing and packaging. When you drink a bottle of their wine, you are literally imbibing their efforts, all of which they are intensely proud. Brian Benson Cellars, his son's project, so successful it has grown out of Dark Star, and moved to Denner Vineyard nearby. Nicole is at Cal Poly studying pre-dentistry...and Norm's passion for producing stellar red wines or "Dark Stars" continues unabated.

DARK STAR CELLARS
2985 Anderson Road
Paso Robles, CA 93446
Phone: 805-237-2389
Fax: 805-237-2589
Web: www.darkstarcellars.com
Hours: Friday-Sunday 10:30 a.m.- 5 p.m.
Case production: 4,000

DENNER VINEYARDS & WINERY

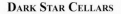

Chances are you have seen various forms of these machines at construction sites; the "Ditch Witch" trenching machines (most wiring, piping, telecommunications, sewer lines, and all underground utilities are increasingly being placed underground these days) trenching the soil for all this underground "stuff" so necessary to our modern lifestyle patterns.

Ron and Marilyn Denner successfully owned and operated several Ditch Witch dealerships in the western USA; Colorado, Idaho, and Utah. They also had an ultimate fantasy of returning to California to become gentlemen vignerons in the truest sense of the word. For years they searched for the perfect spot, literally dozens of visits to Napa and Paso Robles, and eventually found 126 acres of an historic dry farmed barley ranch (owned by the Wiebe family for generations) on Willow Creek Road, that fit the bill. They acquired this and the adjacent property of 30 acres by 2000. Now, just in time for the 2005 harvest, their vision of a superb vineyard, and winery, has come to fruition.

It is the Cinderella slipper of vineyard sites: hilly, calcareous, well-drained soils, cooling breezes from the Templeton Gap, and proximity to the Pacific Ocean. They carefully laid out twenty-five separate blocks taking into account specific mesoclimates within each site, and matching rootstocks and clones to each block.

Photo 26: Vineyards planted to Rhone (Syrah and Grenache) and Zinfandel vines in the Willow Creek district (16 May 2000).

YOUNG PASO VINES

Spanish, Mexican, Italian, Swiss,
Hungarian and French; all have
their interests focused on these wine lands.

Vines will yield the future liquid to be
Distributed and consumed by a
Thirsty international clientele.

Miocene age rocks were laid down as
Ocean sediment, which engulfed the fossils of
Whales, seals, fishes and porpoises.

Famous Monterey Formation rocks decompose into
Soils rich in alkalinity & carbonates, which have
Twin counterparts in French Bordeaux and Rhone.

Worlds connected by culture and geology.

Photos 27-28: Vineyard planting with green inter-row cover crops at Halter Ranch Vineyards (above; 20 December 2001); HMR Vineyard (below; 11 January 2002).

The Denners are dedicated to sustainable farming; no pesticides, all chemicals kept to a very low amounts, permanent cover crops, and deficit irrigation. Their goal is to achieve full organic certification by 2007; they are members of the Central Coast Vineyard Team, a sustainable farming consortium. They are fully invested into the Rhone sensibility: Grenache, Syrah, Mourvedre...and grow the major Bordeaux varietals, and of course, Zinfandel.

The state of the art, gravity flow winery, designed by the well-known architect John Robert Mitchell, was completed in time for the 2005 harvest. It fits seamlessly into the rolling hills, and is one of the few of its kind in California. Gorgeous, to say the least, and a larger, even more beautiful tasting room and events center was opened 2007.

Their son, Brian, is the winemaker. After graduating with a degree in anthropology from Connecticut College, and a year as a ski bum in the Colorado Rockies (actually snowboarding, for you purists out there) a change was beginning to brew. He caught the wine bug, as did many of us, from working at restaurant with a great wine selection. A lot of reading and tastings later, he came to Paso Robles. Brian toiled as a cellar rat at Peachy Canyon Winery during the week, and then assisted on weekends at Dark Star Winery with winemaking and in the tasting room. Still he felt there was more to learn, so he earned a second degree in enology at Fresno State.

With the winds of fortune blowing at his back, he landed a position at the vaunted Williams Selyem Winery, a world-class Pinot Noir producer in the Russian River Valley. For three years, he honed his skills as cellar master, and made wines for the Kingston Family Vineyards in Casablanca, Chile. In 2004, after ten harvests in California and Chile, he took over as winemaker for Denner.

His wines are expressive and typically rich in fruit, Paso Robles style. They are made in classic, noninterventionist style. Given the sellout of their 500-customer limit "Comus (Wine) Club" (named after the Greek god of revelry), Denner has hit the sweet note with their target clientele.

Fruit arrives at the winery, is destemmed and gently moved into fermentation vats via gravity flow. Starting with indigenous yeast fermentation, the caps are punched down by hand, and after maceration are slowly, carefully moved into mostly French oak in the lower barrel cellar. Here again, movement and manipulation is kept to a minimum to preserve the keenly balanced aromatic and fruit flavors. Current releases include a Syrah, and a Viognier/Roussanne blend called Theresa, a southern Rhone styled Grenache/Syrah/Mourvedre blend appropriately called Ditch Digger, a Viognier and, unusually, a late harvest Roussanne.

This is a state of the art, world-class chapter in the Paso Robles story. If you can get on the waiting list for their wines, by all means do so now. It's not going to get any easier to get their limited quantity of wines until production levels increase.

DENNER VINEYARDS & WINERY
5414 Vineyard Drive
Paso Robles, CA 93446
Phone: 805-239-4287
Fax: 805-239-0154
Web: www.dennervineyards.com
Hours: Friday-Sunday 11 a.m.-5 p.m.
        and by appointment

DOVER CANYON WINERY

Situated just six miles east of the Pacific Ocean, in a rustic, beautifully preserved piece of an old walnut orchard, is Dan Panico and Mary Baker's 1921 farmhouse and winery. His comments about his venture are taciturn, "Mountain Grown fruit. Shale. Limestone. Non Irrigated. Six miles from the Pacific."

That does indeed put it in a nutshell, no pun intended here. Dan has been making wine in Paso Robles for close to twenty years, and starting his own production in 1992, while he was winemaker at Eberle Winery. His first wines came from fruit sourced from Norman Vineyards on the westside of Paso Robles. He focuses on dry-farmed, head trained Zinfandel.

The calcareous shale, mudstone and limestone he referred to is a strip of calcareous sedimentary rock of the Monterey

Formation, Sandholdt member, that is located along Vineyard Drive, and is home to many of the areas' finest vineyards. Dan is convinced it is the best region for Zinfandel and Rhone varieties. He produces small lots of wine, many vineyard-designated, from this region. He focuses on using fruit from his own estate vineyard, as well as from the Templeton Gap and Adelaida Hills districts. Their own property, part of the original Rancho Paso de Robles, is a registered wildlife habitat, recognized for sustainable and conscientious planning and execution. The borders remain wild, and are home to a roster of wild life; black-tailed deer, owls, hawks, bobcats and even a cougar with young cubs!

All the wines are completely hand crafted, in a Burgundian style...small half-ton fermenters, hand punched, and gravity fed to barrels. He barrel ferments his white wines. No filtration for the reds. He produces close to half a dozen single vineyard wines at this writing; tiny quantities of Viognier from Hansen Vineyard, in the Templeton-El Pomar mesoclimate, Roussanne from Starr Ranch in the Adelaida Hills district, and a white Rhone blend of the two grapes called White Bone.

His Cujo Zinfandel, produced since 1997, sells out every year, is made with new yeast strains that can deal with high Brix fruit, and spends about 16 months in neutral French and American oak barrels. For another bottling, he sources fruit from the Benito Dusi vineyard, now over 80 years old and planted (unusually for very old Paso Robles vineyards) entirely to Zinfandel. There is a reserve Zinfandel, made with his own estate dry farmed, head trained fruit along with fruit from the Dusi Ranch and Rancho Verano.

His success with Red Rhone varietals is apparent with the immediate sellout of the Syrahs and Rhone blends upon release. He sources Syrah from the Jacobsen's Doce Robles property for his Jimmy's Vineyard and produces a Starr Ranch bottling as well. There is a classically styled Southern Rhone blend called Alto Pomar, a Grenache/Mourvedre/Syrah that is amazing. Bone Lore is his Cabernet/Malbec/Syrah blend...

As if that was not sufficient, in addition to a unfined, unfiltered Cabernet Sauvignon from the Hansen Vineyard, he produces a Bordeaux Blend called Ménage. This wine is a selection from some fascinating vineyard and barrel lots. Cabernet Franc from the historic De Rose Vineyard on an earthquake fault line in Santa Cruz ("practically Pleistocene"), Petite Verdot from Fralich, Cabernet Sauvignon and Merlot from Cougar Ridge Vineyard at 1,600-foot elevation in northwest Paso Robles. There is a Zinfandel Port as well. Whew. These wines are some of the best in the area, with many labels graced by a painting by Ed Smith, vineyard partner and artist, of Dan's beloved, now-departed St. Bernard, Blue.

Hey, you cannot do all of this alone. Dan's muse, Mary Baker is the other half of the operation...running the winery business end, managing the wine club, tasting room and doing the marketing too. She is equally enthusiastic, having just retired as a director in the Paso Robles Wine Country Alliance, where she helped reorganize and reposition this enthusiastic cooperative marketing venture. She also is regional correspondent and editor for Appellation America for the Paso Robles AVA, a published author of several cookbooks, and a self-confessed black thumb. They live with their springer spaniel, Rebel Rose, heir apparent to Blue, in a phenomenal 1920's farm house. Go see them...it is a busy family-run operation with fabulous wines...and their wine blog is one of the best on the international wine web.

DOVER CANYON WINERY
4520 Vineyard Drive
Paso Robles, CA 93446
Phone: 805-237-0101
Fax: 805-237-9191
Web: www.dovercanyon.com
Email: dovercanyon@tcsn.net
Hours: Thursday-Sunday 11 a.m.- 5 p.m.
Case production: 2,500

VINEYARD DRIVE

Wildlands of California's Central Coast,
Converted to domiciles for humans & their crops.

Geology and soil from the Monterey Formation,
Sandholdt member; calcareous sedimentary strata.

Introduced alien vine scions from the old world of
French Bordeaux and Rhone graft to American
Vine rootstocks and find a home.

Vine lands flourishing due to climate and subsurface
Aquifer water that sustains plant growth in droughts.

Feral boar, black-tailed deer, wild turkey, cougar,
Bear, bobcat and coyote electrically
Fenced from the newly planted rows.

Steep rocky slopes, too rugged to cultivate, remain as
Dwelling places for the indigenous plants & animals.

Lone oak tree stands as a sentinel perch for ravens,
Red-tailed hawks, eagles and night predator owls.

(Photo 29: Vineyard landscape along Vineyard Drive: 16 June 2000)

Dunning Vineyards Estate Winery & Country Inn

Robert Dunning got into the wine business from humble beginnings as a home winemaker, where he made wines from local fruit for 15 years. As is a common occurrence, enough good fruit was getting difficult to find. His passion for winemaking was growing, but grape supplies were shrinking.

In 1996, prodded by friends, he entered his wines into southern California's Orange County Fair, one of the biggest and best wine competitions in the country. His Merlot won a gold, and his Cabernet and Chardonnay each took bronze medals. He then established his winery on an 80-acre plot that his family had owned since the 1960's…and with sibling partners, John and Barbara, began planting Cabernet Sauvignon, Merlot, Cabernet Franc, and Chardonnay in 1991. They continue to fine tune the vineyard to this day, planting Zinfandel and Syrah in 2001.

He took classes at UC Davis, but feels that "winemaking really comes from the fields…," where getting the right vine in the right location is paramount. He does much of the work himself, pruning, grape selection and winemaking. Viticulture trellising at Dunning Vineyards is a mixture of VSP (vertical shoot positioning), Geneva double curtain, head pruning and dry farmed, always seeking a balanced and healthy grape vine environment.

The winemaking remains traditional, with small lot fermentation, hand punch downs, and lots of French oak barrels. They produce elegant, rich estate bottled Chardonnay, Syrah, Merlot, and Cabernet. There is the local phenomenon of a house red blend here, too, known as Vin de Casa, a delightfully rich and complex blend of Cabernet Franc, Zinfandel and Syrah, that spends 16 months in French oak, and gets six months of bottle age before release. They are at about 1,500 cases total production now, and don't plan on getting bigger than 3,000 cases…it would take Bob away from his visitors.

There is a calm, pastoral feel to the setting here at Dunning. Bob, Joanne, and their handsome son, Garrett, seem very much at home with the raccoons and the century-old oak trees on the property. They also have a 100-year-old historic farmhouse (represented on the label of all the Estate bottlings) that they have morphed into a gorgeous Country Inn, available for the folks who want a taste of real Paso Robles ambience and charm. The wines are great, the Dunnings are fantastic people…this is a beautiful place to visit, drink up and enjoy the rural native oak forests of Paso.

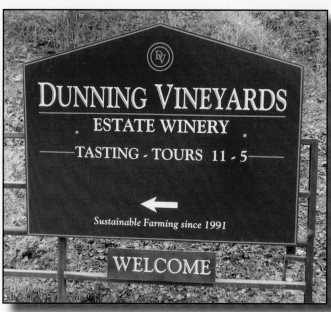

Dunning Vineyards Estate Winery & Country Inn
1953 Niderer Road
Paso Robles, CA 93446
Phone and Fax: 805-238-4763
Web: www.dunningvineyards.com
Hours: Thursday-Monday 11 a.m.-5 p.m.
and by appointment
Case production: 1,500 cases

Four Vines Winery

They call it the Four Vines incident. Christian Tietje (pronounced "tee-gee"), passionate oenophile, and all around nice guy, moves from young chef-dom in Boston, to San Francisco, to pursue his passion for winemaking. He falls promptly and completely for old vine Zinfandels, and relocated to Paso Robles (no mystery here…)

Then, in 1994, he met Susan Mahler, a beautiful blond pilot/scientist with a degree in earth science. Partnership quickly ensued, with Ms. Mahler taking the helm in the vineyards. Four Vines was bonded in 1996… at some point, right around the 15,000 case level, they decided to bring in another hand. Bill Grant, a childhood friend with a successful, entrepreneurial business head came on board. Prompt growth to about 38,000 cases ensued, and continues to the present day.

Christian has an artistic viewpoint about wine, and sources outside of Paso Robles come into play; Santa Barbara for their unoaked "Naked" Chardonnay, Amador foothills (some amazing old vine fruit from the Grand Pere Vineyard goes into their "Maverick" Zinfandel) and Sonoma County as well. That said, close to 70 percent of their production is Paso Robles, and they've got some fantastic examples of the big, lush, red wines that are Paso Robles' calling card.

He is largely self-taught, crafting several vineyard specific Zinfandels, some from Paso Robles, some from old vines in Amador County, north of Paso Robles, and some from (gasp!) Sonoma County. Most of the production of small lot reds is Paso Robles appellation, but their "Naked" Chardonnay has gotten a lot of attention nationally. There is Anarchy, a Syrah, Zin, Mourvedre blend, and the Heretic, a lush, brawny Petite Sirah. There is also the Peasant, a red Rhone Blend comprised of Syrah, Mourvedre, Grenache and, interestingly for Paso Robles, Counoise (one of the lesser known red Rhone varietals), all sourced from Paso westside vineyards.

Tempranillo pops up, too, with "Loco" from the Tres Cajones (look that up in your Spanish dictionary) vineyard in Templeton, blended with Syrah and Grenache. The Biker Zinfandel rounds out the Paso Robles offerings, and is sourced from Preston and Dusi vineyards. They also produce a Zinfandel/Syrah port style wine that spends 20 months in small French oak barriques.

Their tasting room is small, but lots of fun, the wines are fantastic, and they have just completed a new winery on North River Road. Pretty good for a recovering punk rocker...

FOUR VINES WINERY
3750 Highway 46 West
Templeton, CA 93465
Phone: 805-237-0055
Fax: 805-227-0863
Web: www.fourvines.com
Info: info@fourvines.com
Hours: Summer-Daily 11 a.m.-6 p.m.
Winter-Daily 11 a.m.-5 p.m.
Case production: 38,000

GREY WOLF VINEYARDS & CELLARS

The Barton's story starts in Steamboat Springs, Colorado, where a stint as restaurant owners (and wine buyers) gave them the wine bug. Upon moving back to their home state, they started checking out Wine Country...first in Napa then on down to Paso Robles. Joe Barton says that "...when you love great wine and enjoy it with nice people, Paso Robles is the place to be..."

In August of 1994, Grey Wolf was established. Joe and Shirlene Barton assumed full ownership in 1996. It is a family run business, like so many others in Paso Robles. They renovated a great 60-year-old farmhouse on the West side for their tasting room. It is a beautiful, pastoral location with rolling hills and pastures, and used to be called Green Valley. Joe Barton, Sr. sadly passed away in 1998, but the family legacy is still going strong.

Joe Jr., his son and a Cal Poly, San Luis Obispo, fruit science graduate, is making a remarkable assortment of varietals. He is using nearly all Paso Robles grapes, as well as their estate-grown grapes from their 11-acre property. He uses small, open-top fermenters, hand punching, and French, American and Hungarian oak barrels. There are 16 different wines on the current roster with a rich focus on Zinfandel, with several designates headed by their Barton Family Reserve.

Joe is making some fabulous Rhone varietals; red, white and blends...Roussanne/Marsanne blends, a curiously named Chianti Cuvee without a drop of Sangiovese; it is free-run Syrah, Zinfandel and Petite Sirah. There is a Barbera, named the Nomad, a Viognier, and another Syrah named The Predator. The Alpha designation is for their Bordeaux blend of Cabernet/Petite Verdot/Merlot that spends 28 months in oak barrels, joined by Lineage, also a Bordeaux blend, but less reliant on Cabernet Sauvignon. Lone Wolf Red, a blend of Cabernet Sauvignon, Syrah, Mourvedre, and Merlot, is always sold out and rounds out the pack.

Joe's fruit sources are impeccable: the Evelyn Estate, the Jacobsen's, Cerro Prieto, Mélange du Rhone, and Davis...all in the Templeton Gap, Laird on the Estrella Bench, Longshot in Creston and Atascadero in Paradise Valley. He is making full use of the palette of great fruit available to him. These wines are worth seeking out...definitely a leader of the pack in Paso Robles.

GREY WOLF VINEYARDS & CELLARS
2174 Highway 46 West
Paso Robles, CA 93446
Phone: 805-237-0771
Fax: 805-237-9866
Web: www.greywolfcellars.com
Email: greywolf@tcsn.net
Hours: Daily 11 a.m.-5:30 p.m.
Case production: 3,000

L'AVENTURE WINERY AND STEPHAN VINEYARD

Sometimes, when one speaks to a winery principal (winemaker or owner) from the Bordeaux region, there is an almost wistful tone to their voice when they discuss the variety of grapes that, in the USA, we are free to grow anywhere we see fit. You see, in France, indeed much of the European Union (EU), and very much to the chagrin of many a new generation winery owners, the grapes you grow and wines you produce, are legally determined by where your vineyard is located. In Bordeaux, only certain wine grape varieties are allowed to produce wine with the Bordeaux label. They are mostly centered on what are now known as the Bordeaux varietals: Cabernet Sauvignon, Merlot, Cabernet Franc, Petite Verdot, and Malbec for the reds, Sauvignon Blanc, Semillon and Muscadelle for the Whites. If you want to grow Syrah and sell it as Bordeaux, too bad. Chardonnay? Go to Burgundy. The rules are ironclad, and even the smallest changes in the regulations may take decades, if they happen at all.

So when Stephan Asseo came to Paso Robles in the mid-1990's, he was immediately smitten by the unique, varied soils and Mediterranean climate. Coming from his winery, Domaine de Courteillac in the Entre Deux Mers region of Bordeaux, he wanted to be able to experiment with some different non-Bordeaux varietals. He and his wife, Beatrice, purchased 127 acres in 1997 and founded Stephan Vineyard.

His winery operation in California is called L'Aventure, mirroring his dream of blending old world passion for wine and new world technology. His modern winery and superb winemaking, combined with keen attention to the expression of *terroir* in the vineyards, have created wines that have garnered immediate attention from the international wine press.

His plantings of Bordeaux varietals, Cabernet Sauvignon and Petite Verdot, are accompanied by the Rhone selections of Syrah, Grenache, Mourvedre, Roussanne and Viognier. They are all planted at very high density, on mostly south and west slope aspects of calcareous (calcium carbonate marine shells and clay) and siliceous (quartz-rich sedimentary rock) sedimentary rocks, weathering into both alkaline and acid soils (see Figures 16-17).

Meticulous rootstock selection, custom individual row irrigation regimes, Guyot Double Cordon/vertical trellising, regular crop thinning and disking of the bedrock and soil to stimulate deep root growth are crucial to maintaining the very low yields of less than two tons per acre. Stephan harvests with phenolic maturity in mind as well as proper sugar levels and acidity in the grapes. All hand sorted, the reds are 100% de-stemmed and crushed, all lots are fermented separately. He also utilizes pre-fermentation cold soak and micro-oxygenation techniques. He employs 100% new French oak barrels (Troncais and Foret de Jupile) for the fermentations, and

Photos 30-31: Two photos of L'Aventure Winery and Stephan Vineyard taken from the eastern margin of the vineyards toward the northwest. Above; photo taken on 21 August 2006 during period of full vegetative growth. Below; photo taken on 30 March 2007 during early bud break and little vegetative vine growth. These vineyard soils are underlain by both calcareous (alkaline pH) and siliceous (acid pH) Monterey Formation shale and mudstone. See the interpretive soil maps on Figures 16-17 (Rice et al, 2007).

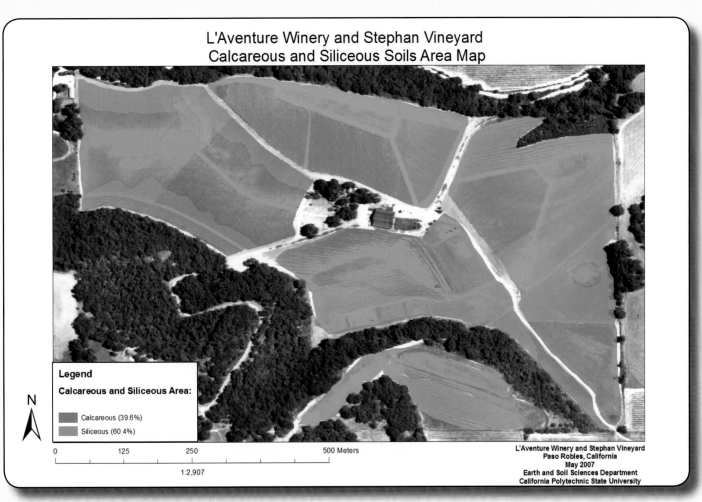

Figures 16-17: Interpretive Soil Maps for L'Aventure Winery and Stephan's Vineyard (Rice et al., 2007).

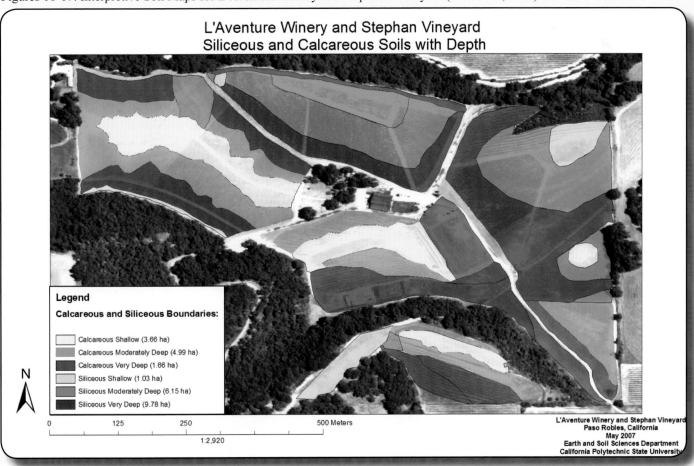

ages his wines for eight to 16 months, depending on the vintage. The core wines are the Estate Cuvee (Cabernet Sauvignon, Syrah and Petite Verdot blends), The Cote-a-Cote (Grenache, Syrah and Mourvedre blends), an Estate Cabernet Sauvignon, an Estate Syrah, and Optimus, his signature blend of 60+% Syrah, with Cabernet and a dash of Zinfandel and/or Petit Verdot. There are limited quantities of Zinfandel, Roussanne, and the Stephan Ridge wines. These complex, unique wines are delicious on release, but also are built for aging, and absolutely worth seeking out.

L'Aventure's adage of "where Bordeaux meets the Rhone" is coming to fruition in a spectacular way, selling out all Stephan's wines and gaining some well-deserved and high profile critical acclaim in an increasingly crowded field. There are a few proud vignerons who insist on the "Best" from vineyard to barrel to bottle, and Stephan Asseo is one of them.

L'AVENTURE WINERY AND STEPHAN VINEYARDS
2815 Live Oak
Paso Robles, CA 93446
Phone: 805-227-1588
Fax: 805-227-6988
Web: www.laventurewinery.com
Email: stephanwines@tcsn.net
Hours: Thursday-Sunday 11 a.m.-4 p.m.
          Everyday by appointment
Case production: 6,000

LINNE CALODO CELLARS

It has a kind of lilting, feminine ring to it, doesn't it? Almost like a country singer's name. But it is all about dirt; oops, soil! Linne Calodo is the name for two soils that are the foundation for the Trevisan family's vineyard along Vineyard Drive...they are two USDA soil series names used for soils weathered from calcareous shale and mudstone, from which they are fortunate to grow their grapes. Calodo's taxonomic family class name: loamy, mixed, superactive, thermic, shallow Calcic Haploxerolls. Whew! Calodo soils had better grow great wine grapes, because you are going to confuse most lay people with that Latin-derived taxonomic soil name.

Another distinction is the slopes on which these grape vines are grown...generally from 15 to 75 percent, which means fantastic drainage, and necessitated the enlistment of the USDA to help with soil erosion problems. These are some steep, hilly vineyards...the vines struggle mightily; yields are low and the grape flavors are very, very concentrated.

Matt Trevisan comes to the wine business from his biochemistry days at Cal Poly, SLO. He cut his teeth with Steve Glossner at Justin Winery in the mid-1990's, and then at Wild Horse Winery. Not a bad start...Matt originally collaborated with Justin Smith in his early (late 1990's) days, but Smith left in 2002 to start up Saxum Winery. Now, Matt is sourcing fruit from first class vineyards such as Denner Vineyards, Lock Ranch, Kiler Canyon, and the famous old vine Zinfandels from Elmer and May Cherry. They are planting an estate vineyard on their 77-acre Oakdale Road and Vineyard Drive ranch, where he and his wife, Maureen, and two daughters have lived since 2000.

Sustainability is a key concept, which will drive the design and planting of his estate vineyards (see photo to the right of new vineyard plantings along Vineyard Drive; 24 July 2007). They are uncompromising with their sustainable viticultural practices. Matt's winemaking philosophy is geared to maximum expression of distinctive *terroir*, focusing on red Rhone Blends, Zinfandel, and Viognier. Grapes are gently de-stemmed and crushed, with a significant percentage of whole berries left intact...small, open topped fermenters, hand punched down and 8-20 months of aging in French and American oak, unfined and unfiltered. Matt's real joy is the blending process, which he feels is critical to the true expression of their *terroir*.

Matt makes several Rhone blends: Nemesis, mostly Syrah, with a little Mourvedre and Grenache, Sticks and Stones, a southern Rhone style based on Grenache, but Syrah and Mourvedre added for structure and mid-palate. There is Rising Tides, Slacker, and Sticks and Stones, all different variations on the Grenache/Syrah/Mourvedre theme. His Zinfandel based wines, Cherry Red, Problem Child, The Outsider, and Leona's, all sport the addition of Syrah and Mourvedre. Tiny quantities of everything, with his wines always selling out. The lone white wine in the lineup is aptly named The Contrarian, a blend of Roussanne and Viognier.

These are patently gorgeous wines, and we can't wait for the estate fruit to show up, as their wine club is full, and we hope for more of these expressions of his superb Paso Robles *terroir* wines.

LINNE CALODO CELLARS
3845 Oakdale Road
Paso Robles, CA 93446
Phone: 805-227-0797
Fax: 805-227-4868
Web: www.linnecalodo.com
Email: info@linnecalodo.com
Hours: Friday-Sunday 11 a.m.-5 p.m.
and by appointment
Case production: 800

MASTANTUONO WINERY (NOW DONATI FAMILY VINEYARD)

During the writing of this book, Mastantuono Winery was sold in early 2007 to Donati Family Vineyard. However, we decided to not to remove this story of an early pioneering winemaker. Here's the original story with several tense revisions...there must be something about Paso Robles that Italians love. These beautiful hills draw them like bees to flowers on warm summer mornings. Pasquale Mastantuono is a perfect example, his ancestral viticultural roots going back four generations to Southern Italy, near Naples. By his teens, he was helping his grandfathers make wine in their home in Detroit, Michigan.

He grew up and moved to Los Angeles, where he became a renowned custom furniture maker to the likes of Sammy Davis Jr., and Elvis Presley.

He went to more than a few Hollywood soirees, and was disappointed with the wines they served. He decided to make wine at his villa in Topanga Canyon, and the rest is history.

It wasn't too big of a jump to his purchase of 65 acres on the westside of Paso, near Templeton. He planted 17 acres of Zinfandel in 1977. His first release, a Zinfandel from Dante Dusi's old vines that same year, yielded such fantastic results that he became known as the "Zinman," a moniker that remains to this day.

He designed and built the winery, tasting room and home, which have a decidedly European feel to them, "an English hunting lodge outside and Irish Pub inside," as was aptly stated on their charming web site.

He made about 10,000 cases, with Zinfandel holding center court. They then expanded into Italian varietals: Barbera, Sangiovese, and Muscat Canelli as well. He produced a red wine called Carminello, from the hybrid Carmine grape, a cross between Cabernet Sauvignon, Merlot and Carignane. He also produced a Syrah, and port styled wines as well. Rounding out the selection was a Chardonnay, White Zin and a couple of sparkling wines...this is still a charming place to visit with the Donati Winery. Pasquale, the wines and his famous wild boar feasts will be missed by his many fans.

MIDNIGHT CELLARS

Bob and Mary Jane Hartenburger came to California wine country on vacations from the wilds of Chicago. Accompanied by their son Rich and daughter-in-law Michele, Bob took his son up on the loaded suggestion that he start a winery and let him and Michele run it. They scouted properties in all the major areas; Napa, Sonoma, and Temecula, finally settling on the Central Coast, where the cost of land wasn't so dear (yet) and the reputation for quality wine grapes was becoming well known. Their estate was originally a barley and hay farm. They converted the horse barn to a winery, and planted thirty acres of Cabernet Sauvignon, Merlot, Zinfandel, Petite Verdot and Chardonnay.

Bonding came through in 1996, and they released their first wines that year. Now, 11 years later, the horse barn has been replaced by a 10,000 square foot building next to the tasting room on Anderson Road. The new winery has a laboratory, barrel room and a 25,000-gallon capacity stainless steel tank room. Rich, is a self-taught winemaker, and makes all the wines.

They estate bottle about 65 % of their production, concentrating on superb red varietals that they grow, and source from selected Paso Robles vineyards. Rich uses a combination of American, French and Eastern European oak, about 25% new every year. They make a beautiful Pinot Noir from the Latchford Ranch across from the winery, planted to Burgundian clones and aged in French oak. There is a Cabernet Sauvignon, "Nebula" , a reserve Zinfandel/Syrah blend named "Gemini", selected from their best barrels, and "Full Moon Red" the regular bottling. Rich makes an estate bottled Merlot, and a "Nocturne" Syrah sourced from three distinct east and westside vineyards. "Mare Nectaris" is their classic Bordeaux blend of all five varietals: Cabernet Sauvignon, Merlot, Petite Verdot, from their estate, blended with Cabernet Franc and Malbec.

It wouldn't be a complete Paso Robles story without an estate bottled Late Harvest Zinfandel, and its complement, a non-vintage "Gemini" Port comprised of about 70% Syrah and 30% Zinfandel. Total production is about 7,000 cases...and with high scores from the wine writers and gold medals being awarded from national and international competitions, the Hartenburger's have achieved..."Midnight Magic, Stellar Wines."

MIDNIGHT CELLARS
2925 Anderson Road
Paso Robles, CA 93446

Phone: 805-239-8904
Fax: 805-239-3289
Web: www.midnightcellars.com
Email: info@midnightcellars.com
Hours: Daily 10 a.m.-5:30 p.m.
Case production: 7,000

OPOLO VINEYARDS

Dalmatia, on the Croatian coast, would seem an unlikely place to source a winery name from, but that is what Dave Nichols and Rick Quinn did. Opolo is the term for a deep rose wine. And, in a bow to their eastern European heritage, Opolo Vineyards and winery were christened in 2001.

They had been neighbors in Camarillo since 1996, when Rick began planting vines throughout Paso Robles. His love of winemaking stretches back to his youth in Minnesota, in a multicultural neighborhood of family winemakers, making wine from purchased fruit. He moved to California in 1979, and began to source grapes from the Paso Robles region. When supplies tightened up in the mid-nineties, as wineries made more and more wine under their own labels, he bought a westside property without even looking at it first, and began planting vines.

They now own over 300 acres total, about 225 on the eastside and 80 on the westside. The original plantings were Zinfandel, Pinot Noir, Merlot, Cabernets Sauvignon and Franc, Sangiovese and Chardonnay. In 1997, they planted Syrah, more Cabernet Sauvignon, Merlot and Muscat Canelli. Some recent grafting over to Roussanne, Viognier, Tempranillo and Barbera has occurred as well.

The winery itself is a restored old tractor building, created for the purpose of housing machinery for the vineyards...the original intent was to produce fruit for sale to wineries like Hess Collection, Wild Horse and Fetzer, which still is the case. But a certain percentage goes to the Opolo label, with spectacular results. Ninety percent of their production is Paso Robles, 70% of their varietal production is estate bottled.

Dave and Rick make all the wines, assisted by Robert Nadeau. Rick explains that "it's kind of winemaking by committee...we love to use oak, but match the wine to the barrel..." French, American, and Hungarian oak barrels all find their way onto their diverse winemaking spice rack.

They produce close to twenty varietals, concentrating on the classic red Paso Robles varietals of Zinfandel, Syrah, Petite Sirah, Rhone and Bordeaux style blends. They have released the inaugural vintage of Fusion, a 50/50 Cabernet Sauvignon/Syrah blend. Rounding out the white side of their portfolio are small lots of Pinot Grigio, Viognier, Roussanne and Muscat Canelli.

There is an open, easy feel to the winery and tasting room, which sits atop a beautiful 70-acre mountain top site. The current 7,000 square foot facility is slated for expansion to 12,000 square feet, and plans are afoot for an eight-room bed and breakfast, too. It sounds like Opolo's success story keeps getting better.

OPOLO VINEYARDS
7110 Vineyard Drive
Paso Robles, CA 93446
Phone: 805-238-9593
Fax: 805-371-0102
Web: www.opolo.com
Hours: Daily 10 a.m.-5 p.m.
Case production: 3,000

PIPESTONE VINEYARDS

Jeff Pipes is another transplant success story, with a couple of unusual twists...he (also a former geologist and lawyer) left 22 years of education, three degrees and a flourishing environmental law practice in Minnesota to come to Paso Robles to live his dream of planting a vineyard and making wine. His Hong Kong born wife, Florence Wong (they met as undergraduates in Minnesota) has had some unique influences on the operations as well. Pipestone may be one of the very few, if not the only, USA vineyard and winery to be laid out according to the principals of Feng Shui. They have a Feng Shui master in Hong Kong, a certain Mr. Wong (no relation to Florence) that keeps all their energies balanced and flowing harmoniously.

This ancient Chinese practice of harnessing the natural earth energies to enhance all aspects of life would seem to have a natural application in a vineyard. If you have ever read a Farmer's Almanac, or are familiar with the principles of biodynamics, this will be familiar territory.

They settled in 1997 on a 30-acre westside parcel with their young daughter, Grace (now accompanied by a younger sister, Gwen), and began the transformation of Pipestone. Doing all but the vineyard planting themselves, they also have a beautiful organic garden and raise a host of farm poultry. The vines are hand tended, with a new barn to house draft horses, who will shoulder the task of tilling the vineyards, turning the varied cover crops between the rows into the soil each spring, and avoiding the soil compaction that comes along with heavy equipment use.

They really are trying to avoid any harmful input into their (and our) precious earth; Pipestone has been organically farmed since 2000. They have left much of the property largely unfenced, to allow deer to pass through, and have left the riparian areas surrounding the vineyards intact, which encourages natural predator and prey cycle that keeps all the complex biodiversity in balance.

Jeff makes all the wines; using a combination of American oak (from his home state of Minnesota) and French oak from the Nevers forests. His interest in wine comes from his grandfather, who came back from his duties in WWI France, having fallen in love with great food and wine. Jeff loved to hang around and watch his grandpa make wines, mostly from fruit and hybrid grapes, and it is now his vocation.

Their current line up is 90% estate bottled, all but a dry farmed, head pruned, own-rooted Zinfandel they source from their Sri Lankan neighbor. The only white wine they make is Viognier, which sees some French oak and sur lie aging. They make a Grenache, and some interesting takes on Rhone varietals: A Grenache/Syrah/Mourvedre blend, a varietal Mourvedre, a gorgeous Syrah and reserve Syrah, and a dry, rich Grenache Rose, which is the only wine of theirs that is partially stainless steel fermented.

They don't want to be any bigger than 2,000 cases...Jeff explains "we want to do it all ourselves, to make a really hand made wine..." and more production would mean less time for the finely detailed, individual attention each wine gets. A wonderful attitude that will serve Pipestone well now and into the future.

PIPESTONE VINEYARDS
2040 Niderer Road
Paso Robles, CA 93446
Phone: 805-227-6385
Fax: 805-227-6383
Hours: Thursday-Monday 11 a.m.- 5 p.m.
Web: www.pipestonevineyards.com
Hours: Thursday-Monday 11 a.m.-5 p.m.
Case production: 2,000 cases

ROTTA WINERY

World War I had not even started in 1908, when Joe Rotta bought a successful Paso Robles vineyard from Frenchman Adolf Siot. Joe in turn sold it to his brother, Clement, who in turn bonded the Rotta winery after Prohibition in the thirties. Now, in this 21st century, this historic property is well along the comeback trail, led by Michael Guibbini, a Rotta grandson who remembers his grandparents working in the vineyards. He was so enamored of the area, he moved here after high school, attended Cal Poly in San Luis Obispo, and settled permanently in Paso Robles.

During a 30-year career as a fire captain for the California Forestry Department, Michael decided to replant the 37-acre old family vineyard. In 1990 he planted 20 acres of Zinfandel (some of which is own-rooted, some is on St. George rootstock) and a little Cabernet Sauvignon and Cab Franc. All are head pruned and dry farmed. Until 2002, their Zinfandel went into a Giubbini vineyard designate for Castoro Cellars. About 10 percent of their production is estate bottled, 95 percent of the production is Paso Robles fruit. They source a little Chardonnay from Monterey. They source a fascinating, single vineyard planted in 1938, in Fresno for Manukka (a Jerez, Spain varietal) for a fortified dessert style wine, made in a genuine solera style, spending two years outdoors in barrels before bottling.

Winemaker Mark Caporale, a UC Davis fermentation science graduate, hails from Napa Valley. His family made Zinfandel for their own use for years, and by the time he was in the ninth grade, he knew what he wanted to do with his life. He is a non-interventionist, to a point, but happily explains that when he came to the Rotta's property "....Mother Nature in the vineyard was just kind of telling me; pick me before the rains come...and just don't screw it up!" He use 24-72 hours of cold soak for

most of the red wines, and about 30% new French and American oak for everything except the Merlot, which sees only French oak barrels. All wines are lightly fined and filtered.

They produce a Muscat Canelli sourced from Paso Robles eastside vineyards, and get some Cabernet Sauvignon from Rainbow's End on the north side of Paso Robles. They have made a real commitment to Cabernet Franc, sourcing the fantastic Boneso Vineyard fruit "...you would swear you are in Bordeaux when you look at the soil...," Caporale explains. They are fast becoming the leader in the Central Coast for this varietal. They also produce a deeply colored, firm, off-dry Rose of Zinfandel.

The historic winery is being remodeled, there is a new crush facility being built on the property, and a new tasting room to boot. We cannot wait to see the new digs, and taste the latest releases from this genuine "real deal" Paso Robles family whose roots run deep into the Paso Robles AVA.

ROTTA WINERY (TASTING ROOM ADDRESS)
3750 Highway 46 West
Templeton, CA 93465
Phone: 805-434-9621
Fax: 805-434-9263
Web: www.rottawinery.com
Email: rottawine@tcsn.net
Hours: Daily 11 a.m.-5 p.m.
Case production: 5,000

"IN WINE ONE BEHOLDS THE HEART OF ANOTHER."
ANONYMOUS

SANTA MARGARITA RANCH (approximately 18,300 acres) and the southern expansion of the Paso Robles AVA (approximately 2,635 acres): cooler region II climate with pronounced maritime and orographic influences, with cold air drainage and ponding, high steep mountain slopes of Santa Lucia Range down to valley floor of incised Salinas River, diverse bedrock in area of many formations, diversity of soil types by bedrock and slope position, with most vineyards on river terraces with deep alluvial soils, oak savannah in the valley floors to chaparral and mixed woodlands on the hillslopes (Elliott-Fisk, 2007).

Google Earth image of the village of Santa Margarita and vineyards to the southeast within the Santa Margarita Ranch (above).

Photos 32-33: Oblique airphotos of selected Santa Margarita Ranch vineyards and adjacent ranch lands (below: 25 March 2003).

"WINE MAKES DAILY LIVING EASIER, LESS HURRIED,
WITH FEWER TENSIONS AND MORE TOLERANCE."
BENJAMIN FRANKLIN

PASO ROBLES ESTRELLA DISTRICT (approximately 66,800 acres): moderate-low region II-III climate, with some maritime influence, largely valley floor topography of Estrella River floodplain and younger to older river terraces across Quaternary alluvium, alluvial soils of diverse ages on flight of terraces, with some of these loamy soils cemented by calcium carbonate, silicates, irons and clays, oak savannah vegetation (Elliott-Fisk, 2007).

BIANCHI WINERY

Their beginnings in the early 1970's were in Kerman in the Central Valley, with Joseph Bianchi making the wines and his son Glenn serving as sales manager. They still operate the Kerman property, but in 2004 opened a new winery in Paso Robles, and are estate bottling from their own Paso Robles fruit about 60 percent of their production in this sparkling new facility. They purchased a 40-acre vineyard in 2000, planted to Cabernet Sauvignon, Zinfandel and Syrah. They are striving for 100% percent estate bottled production.

Their estate vineyard is comprised of four different ranches. The 13-acre Lakeside Ranch planted to Cabernet is based on clay loam. The 19-acre Zen Ranch is mainly sandy loam and planted to Zinfandel. The Creekside and Sunrise Ranches are planted to Syrah and Merlot respectively. Glenn Bianchi is now managing general partner; the winemaking is directed by Tom Lane, a 22-year veteran of wineries such as Navarro and Concannon, with degrees in biology, botany and chemistry, as well as a UC Davis enology graduate.

They source some Sauvignon Blanc from south Monterey County, Pinot Grigio from Arroyo Grande Valley, Petite Sirah from the Paso Robles Peck Ranch, Pinot Noir from York Mountain, and Cabernet Franc from Jim Smoot's Vineyards. When asked what his favorite varietal is, Tom flatly answered "...I don't answer that question...it's like asking a parent who is your favorite child..." They are making 13 different varietals now, about 5,000 cases, with the Heritage (all estate fruit) wines as the flagship, buttressed by the Signature line of small lots of more widely sourced, economically priced wines.

Tom's viewpoint about oak is only to use it where it's most appropriate, as a spicing tool... and that he "likes to taste the grapes." They use 25-35% new French and American oak, utilizing a 1-4 day cold soak for the red wines to extract color and flavor. There is a new program for Refosco, a northeastern Italian red grape that ripens late and produces firm, deeply colored dry wines. This fruit comes from the San Juan Vineyard. Tom gives it a year in French oak, and the result is unusual and interesting. This is a spectacular new destination in Paso Robles, and there is a wonderful new vineyard house available for those in need of a Paso getaway.

BIANCHI WINERY
3380 Branch Road
Paso Robles, CA 93446
Phone: 805-226-9922
Fax: 805-226-8230
Web: www.bianchiwine.com
Email: tr@bianchiwine.com
Hours: Daily 10 a.m.-5 p.m.
Case production: 25,000

EBERLE WINERY

Gary Eberle's impact on Paso Robles story cannot be understated. During the 1950's and 60's, dairy and cattle ranches still predominated; no Paso Robles wineries were really competing on a national or international level. That all changed when Gary arrived in 1972 to begin his seminal vision of world-class wine from the Paso region.

Here is a native of Pittsburgh, Pennsylvania, with an unusual combination of skills; a former defensive lineman at Penn State who completed graduate work in cellular genetics at LSU. During his stint at LSU, after being introduced to classified growth Bordeaux in a professor's cellar, he was immediately smitten. It was a short jump to California, where he convinced the UC Davis Enology Department chair to admit him to the doctoral program without an entrance exam.

During this time at Davis, he realized the enormous potential of the Paso Robles region, where he founded the original Estrella River Winery in the early 1970's. Dreaming of still producing ultra-premium red wines, he released his first "Eberle" wine, a 1979 Cabernet Sauvignon, in 1982. The home estate at Eberle is a 40-acre parcel surrounding the winery, planted with Cabernet Sauvignon, Chardonnay and Muscat Canelli, on well-drained, gravelly loam soil derived from the Paso Robles Formation. Eberle wines are made from grapes coming an assortment of local vineyards. With the old Dry Creek floodplain running through it, the Steinbeck Vineyards, on the south side of Highway 46 East, is planted to Syrah, Zinfandel and Barbera. The Mill Road Vineyard, adjacent to Steinbeck, is planted to Viognier. Lonesome Oak has plantings of Syrah and Grenache Noir. The Remo Belli Vineyard, on the westside, rounds out the picture with its 40-year-old, head pruned, Vinifera-rooted, dry farmed Zinfandel vines.

With wonderful vineyard holdings and partners like this, he has grown to produce 14 varietals and over 25,000 cases of wine. The quest for more space took him underground...where, in 1996, he developed an 8,000 square foot wine cave that provides an ideal environment for aging his red wines (see photo below). This unique facility expanded to 16,000 square feet in 1999, and enables them to use gravity flow transfer methods for production. These barrel rooms are a spectacular venue for the regular public and private events, as well.

The German name "Eberle" translates to "small boar." A bronze replica of a 17th century Porcellino cast by Tacca is at the tasting room entrance, where visitors in search of good luck rub its nose and toss coins in the water below. For over 25 years, Eberle winery has celebrated the Paso Robles *terroir* with their unique facility and luxury wines, a story that is bound to continue and flourish.

EBERLE WINERY
P.O. Box 2459
Paso Robles, CA 93447
Phone: 805-238-9607
Fax: 805-237-0344
Web: www.eberlewinery.com
Email: tastingroom@eberlewinery.com
Hours: Summer, 10 a.m.-6 p.m.;
        Winter, 10 a.m.-5 p.m.
Case production: 25,000

J. Lohr Vineyards & Wines

Jerry Lohr was born and raised on what would now be called an organic farm in South Dakota. This background instilled a real sense of how important soil, climate and slope aspect, (*i.e., terroir*), are to creating outstanding vineyards and wines. Farming is literally, in his blood. After graduating from Stanford University, and a stint in the Air Force as captain at the NASA Ames Research Center, he began his journey of investigation of grape growing regions throughout the state. In 1972, he purchased 280 acres in Arroyo Seco (Monterey); this has since grown to 900 acres. In 1988, he bought property in Paso Robles, currently at 2,000 acres, 1,250 of which are primarily planted to Cabernet Sauvignon, Merlot and other red varietals. Some 750 acres are currently slated for planting as well.

I was privileged to have a personal tour in a truly enormous state of the art computer controlled tractor, piloted by the master himself (see photo to the right). We traversed a new hilltop vineyard site, turning over the dolomite and magnesium applications two feet into the sandy loam soil. Soon the irrigation system will be installed, and then special grasses will be planted, to be plowed under. This painstaking process bolsters the organic component and water retention capacity of this sandy loam soil. Later on, when the vineyard (planted to Cabernet Sauvignon, Merlot and Petite Verdot) is at full production, special J. Lohr pomace will be added to the soil on an ongoing basis. This vineyard, like most of his vineyards, trellised with what he calls a "modified sprawl" will be a great example of his philosophy of growing the components of his different wines within the same vineyard mesoclimates.

J. Lohr's talented winemaking team of Jeff Meier and Dan Shaw produce a truly remarkable range of wines, seeking always to display the lush, mouth filling flavors that are Paso Robles hallmarks. At the entry level, the popular Cypress series, then the Seven Oaks series, from south facing slopes and French clones. There is a Crosspoint Pinot Noir, South Ridge Syrah, marvelous old vine Bramblewood Zinfandel and Los Osos Merlot, and a superb Hilltop Vineyard Cabernet, always one of the popular Paso Robles Cabernets.

His latest endeavor is the Cuvee series, which I was fortunate to taste alongside a number of classed growth Bordeaux at the winery in 2004. These wines are produced in extremely limited quantities, and represent an artistic blending philosophy as exemplified by the classic wines of Paulliac, St. Emilion and Pomerol. Cuvee St. PAU is at least 50 percent Cabernet Sauvignon, with Merlot, Cabernet Franc, Petit Verdot, and Malbec. Cuvee St. E, produced from over 50% Cabernet Franc, along with Merlot, Cabernet Sauvignon and Petite Verdot. The Cuvee POM is over 50% Merlot with Cabernet Franc, Cabernet Sauvignon and Petit Verdot. These stunning wines will challenge the most expensive wines in Bordeaux and California.

J. Lohr Wines have become a very visible example of the level of quality that is possible from Paso Robles (see the photo above showing an aerial view of the J. Lohr wine tasting building and the Hilltop Vineyard located north of the Paso Robles airport; 24 March 2003). Jerry Lohr is a tireless evangelizer for California wines and sustainable agriculture practices. His work goes well beyond the vineyards and wines, including posts as former chair of the Wine Institute, founding member of Wine Vision, and past director and chair of the marketing committee for the Paso Robles Vintners and Growers Association (now called the Paso Robles Wine Country Alliance). Although others have followed in his path, Jerry remains one of the few winery owners who can still be found driving a tractor, his destiny firmly rooted in his vineyards.

J. LOHR VINEYARDS & WINES
6169 Airport Road
Paso Robles, CA 93446
Phone: 805-239-8900
Fax: 805-239-0365
Web: www.jlohr.com
Hours: Daily 10 a.m.-5 p.m.
Case production: 740,000

EOS ESTATE WINERY

This is a truly American rags to riches story. When Frank and Phil Arciero landed in Detroit in 1939, neither of them spoke a word of English. Frank worked as a manual laborer while Phil went to school. Their father, Giovanni, had been in the USA since 1914, but the rest of the family did not come until 1937, starting with the oldest brother, Mike, and finishing with mother Christina in 1948. By then, Frank and Phil had moved to California, and after dreams of Hollywood careers faltered, they worked as laborers, but quickly launched their own concrete business, the Arciero Brothers. This business, which is still very prominent today, was to become one of the largest in California, and is the foundation upon which the winery was built. Frank Sr. and Jr. also developed a passion for car racing; Indy Cars and Baja Style- check out the race car exhibit at the winery!. Phil Arciero is a world-class big game hunter and trophy fisherman, one of the few American sportsmen to have earned the "Big 5."

When the Arciero family, Vern Underwood and Kerry Vix got together in 1986 about starting up a new east side Paso Robles winery, they didn't have far to look. Frank Arciero, Sr. had been buying land in the area since the early 1980's, after building the extremely successful concrete and construction development business with Phil. They had built a magnificent 104,000 square foot winery in 1986, and added a 25,000 square foot high-tech barrel facility in 1998, which was built into a hillside to conserve energy. The 400,000 case capacity winery is styled after the historic Benedictine monastery in Italy, Montecassino, which is near Frank and Phil's family home of Saint Elia Fiumerapido, 129 km south of Rome. All that was needed was the vision and drive to launch a new luxury Paso Robles brand, and EOS was born.

EOS is the Classical Greek mythological name for the Goddess of Dawn, inspired by the pre-dawn harvest of all the fruit grown on their 700+ acres. She was sister to Helios (Sun) and Selene (Moon), as well as mother to the four winds: Boreas, Eurus, Zephyrus and Notus. She also gave birth to Heosphorus and the Stars. EOS was a bit of a trollop, and earned the wrath of Aphrodite by trysting with Ares. One of her lovers was Orion, whose constellation appears on the EOS Estate label today.

Leslie Melendez joined EOS in 1993 and was appointed senior winemaker in 2004. The distinctive, rich, fruit forward style of the EOS wines is her hallmark. The range is remarkable as well, with close to a dozen varietals planted: along with Chardonnay, Cabernet and Merlot, there is Sauvignon Blanc, Zinfandel, and a world class Petite Sirah. The Cupa Grandis ("great barrel" in Latin) wines, their top of the line reserve Petite Sirah and Chardonnay are some of the richest and best wines to come from Paso Robles.

Also grown are Sangiovese, Pinot Grigio, Nebbiolo, Cabernet Franc and Chenin Blanc and more than a little Muscat Cannelli, which goes into their wonderful late harvest "Tears of Dew." There is a terrific Zinfandel Port, too. There is constant refining and expansion of the winery and vineyards, with multiple layers of experimentation; new clones (some very promising Clone 6-Jackson Cabernet, and new Dijon Clone Chardonnay are in the pipeline), and French and American oak barrel regimes are constantly being tested. EOS is an impressive, state of the art winery, with all the bells and whistles of 21st century winemaking, but this focused and talented team brings it all together. The Novella line was introduced in 2001, the wines being made in a very fresh, user-friendly style. Novella

Synergy, a fabulous red blend of Zinfandel, Sangiovese and Petite Sirah, was an immediate hit with wine drinkers and critics alike, and sells out with every vintage. The EOS Visitor Center is a 6,000 square foot Mediterranean marketplace on five and half acres of landscaped rose gardens, fountain and picnic area. It has become a true Paso Robles destination, and is a genuinely difficult place to emerge from empty handed. The wines are consistently some of the finest in the entire appellation. This is a classic Paso Robles story; hard work, vision and real sense of place coming together with great winemaking…a great story that continues to evolve. EOS has just been purchased in summer of 2007 by Blavod Extreme Spirits, who plan to transform it into a Paso Robles showplace, with Kerry Vix and his team staying in their current positions.

EOS ESTATE WINERY

(Tasting room: Highway 46 East,
6 miles east of Hwy. 101)
P.O. Box 1287
Paso Robles, CA 93447
Phone: 805-239-2562
Fax: 805-239-2317
Web: www.eosvintage.com
Email: friends@eosvintage.com
Hours: Daily 10 a.m.-5 p.m.;
        Summer weekends 10 a.m.-6 p.m.
Case production: 160,000

GARRETSON WINE COMPANY

Matt Garretson is a moving target. I had to catch him on his mobile phone; between the winery and home, "...I guess in retrospect it wasn't the smartest thing to do, starting a winery and a family at the same time..." However, start them he did, in 2001, after previous incarnations as a fine wine retailer, wine bar owner, importer, wholesaler, and winery manager.

His love affair with Rhone varietals began with a gift of a bottle of Condrieu, and has blossomed into one of the driving forces for these varietals in Paso Robles. The former "Mr. Viognier" founded the Viognier Guild, which has since morphed into the Hospice du Rhone, the world's largest celebration of Rhone wines. This event in Paso Robles every year draws a global audience of fine wine aficionados, and owes it phenomenal success to a number of different wineries and personalities, but none more so that Matt Garretson.

His wine portfolio, prefaced with "No Cab. No Chard. No Merlot. No Crybabies" is centered on, not surprisingly, Rhone varietals. He does not own a vineyard, but sources grapes almost completely from Paso Robles. They are nationally distributed and in four countries now...with production at its current maximum of 12,000 cases; Matt does not want to grow bigger than that. He has been making wines since 1982, but the last ten years have really refined and defined his style.

He uses cold barrels for his white wines, and has utilized must pre-oxidative techniques. He is "getting away from that now..." but the cold barrels remain, in a huge refrigerated trailer (at about 45 degrees Fahrenheit), and fermentations taking months to complete. The red wines are generally destemmed and he uses a lot of whole berries, new and second/third fill French oak barrels, with some American oak as well. Very little fining or filtration for any of his dry wines. There isn't a formula; one senses that he really listens to what the fruit is trying to tell him.

He wanted to thumb his nose at the stodgy, French-themed names and labels... and the wines have beautiful Gaelic names: Viogniers "The Chumrah" and The Saothar, "The Limoid Cior" Roussanne, and "The Berwyn," a Roussanne Vin Doux named for his son Thomas Berwyn Garretson, born during the 2000 Harvest...Berwyn means "son of the harvest." The wine is fortified with his very rare "Cuvee Ella," an Eaux de vie (grape spirit) and is distinctly sweet. "The Celeidh" is his name for their Syrah/Mourvedre/Grenache rose...named after the Gaelic for celebration. He makes a gorgeous palette of Syrahs, "The Aisling" is the sole 100% Syrah, and it sees more than a little new oak. "The Craic" means celebration. This wine is a classic Northern Rhone style blend of Syrah with a little Viognier, which lifts the aromatics. He uses some whole berry Syrah fruit during fermentation, and French oak. "Mon Amie" is named for his lovely wife, Amie; this San Luis Obispo appellation wine is blended with Viognier. "The Bulladoir" is a huge, inky, extracted wine, named in tribute to his home state's

University of Georgia. His vineyard sources? "We are really excited about fruit from the Templeton Gap area...we are starting to focus more and more here, for Syrah, Viognier and Roussanne. Grenache does well there too..." Canopy and crop management play a big role in achieving more moderate alcohol levels, along with better pH and acid balance. "We are an industry that tends to be knee jerk in reactions: lower yields, leaf pulling, denuding the fruit zone of canopy...I'm realizing that it depends on the varietal, that higher yields can mean better quality in some cases...other than to control mildew issues, you don't want to denude the fruit zone..."

He is particularly proud of his new Late Bottled Viognier, and the new Grenache "The Spainneach." The Viognier is a true Condrieu style, sourced from 20 of his best barrels that he held back, and kept there for 13 months total. He is going forward with this style for his Viognier program. "...I almost gave up on Viognier...there are so many disappointing wines...." His Grenache has been a rewarding challenge "...it has always been a headache...oxidation, tannin management, alcohol levels...to find a Holy Grail that turns you on..."

Matt never met a Rhone varietal he didn't love. He makes a Mourvedre "The Graosta" and another Roussanne called "The Capall Allas" which refers in Gaelic fashion to, well, horse sweat. I'm sure the wine is more appetizing than that, although he does note the need for decanting...hmmm... He has a second "The Berwyn" made from Grenache Blanc...He has titled his reserve flagship wines "The Reliquary" meaning a vessel that holds a sacred object... a fitting name for the very limited production red and white Rhone blends. There is also a "G" Series entry level red and white bottling.

I can't imagine Matt's portfolio or his story remaining static. He is constantly learning, modifying, progressing. He and his lovely wife Amie, handsome and handful sons Jack Francis and Thomas Berwyn, will no doubt be playing to generations of enthralled customers. Go Georgia Orange Bulldogs!

GARRETSON WINE COMPANY
2323 Tuley Court, Suite 110
Paso Robles, CA 93446
Phone: 805-239-2074
Fax: 805-239-2057
Web: www.garretsonwines.com
Email: info@garretsonwines.com
Hours: Daily 11 a.m.-5 p.m.
Case Production: 15,000

LAURA'S VINEYARD

Ray and Pam Derby made Cambria their home in 1995, after a successful career in manufacturing. They purchased the Central Coast Derbyshire Vineyard in 1998, acquiring the 320-acre eastside Paso Robles Laura's Vineyard, in 2001. They then bought the Rozet vineyard in 2006, a prime westside property in the Templeton Gap, which is planted to Rhone varieties. The tasting room on the eastside vineyard was purchased in 2005 as well, and plans are afoot for the new winery there, too...they never had any intention to do anything but grow and sell grapes, but when they purchased the westside property, they saw the opportunity for making and marketing their own estate wines. They will estate bottle their own production from the 2005 vintage on, and will increase production to about 5,000 cases.

Tiffinee Vierra, a Cal Poly SLO graduate, is the winemaker and GM. She came to Laura's in August of 2005, after excellent beginnings at Wild Horse, Edna Valley, Tablas Creek and Four Vines. "I really am excited about all three properties...having these resources is so fantastic...I'm having a great time." When asked for her winemaking philosophy, she replies that "...I don't mean to be redundant, but our story is so unique, each vineyard has such distinct characteristics. I use new oak, I cold soak, I do all these things stylistically but the goal is to emphasize the *terroir*. As a winemaker, I want the wines to represent where the fruit is grown." Her husband, Steve, is a well respected soil scientist and irrigation specialist for Vineyard Professional Services in Templeton and he will surely have some impact on the vineyard management decisions; while the rest of us can enjoy Tiffanee's wines from the estate grapes.

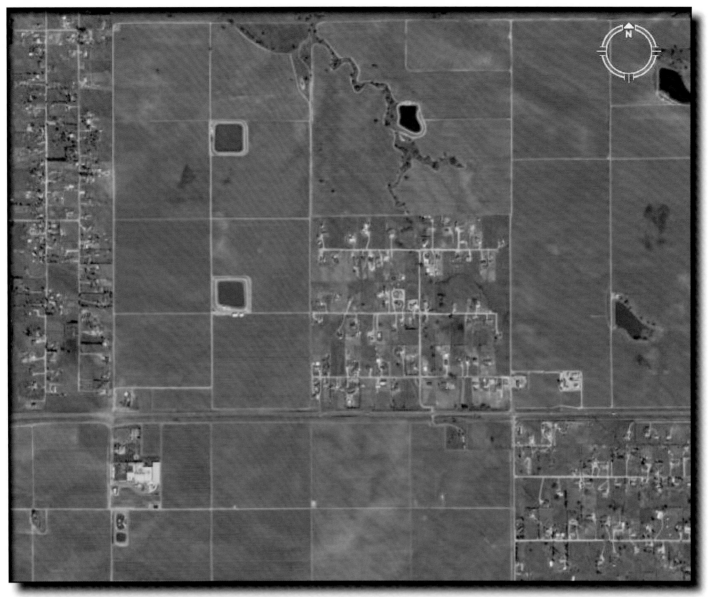

Google Earth image of EOS and Arciero Winery, Laura's Vineyard, and the surrounding lands along Highway 46 East.

Photo 32: Steve Vierra, soil scientist for Vineyard Professional Services, mapping soils at Laura's Vineyard (July 1999).

They've just released a newly designed label for their rose of Syrah. There will be a name change in the future, a new brand to match the Mission San Luis Rey-themed architecture for the winery. "We are really big on embracing the history of California...we wanted to stay away from French and Italian themes. We want to focus on Paso Robles and the history of the area, and make wines that are great examples of the three different properties that we have..."

The new 2007 releases include their Estate Cabernet Sauvignon, a white Rhone Blend, made from Viognier, Roussanne, and Marsanne, and a Central Coast Pinot Noir. There is a rush of new activity at Laura's Vineyard, with more excitement to come in the future.

LAURA'S VINEYARD
5620 Highway 46 East
Paso Robles, CA 93446
Phone: 805-238-6300
Fax: 805-238-6911
Web: www.laurasvineyard.com
Email: info@laurasvineyard.com
Hours: Daily 10 a.m.-5 p.m.
Case Production: 5,000

MARTIN AND WEYRICH WINERY

If you have ever been to Tuscany, the Italianate villa sited directly north of Hwy. 46 east might just take you aback. The resemblance is more than coincidental. Martin and Weyrich Winery (formerly Martin Brothers Winery) is the result of a great deal of hard work and focus by Dave and Mary (Martin) Weyrich, who took over the winery in 1998, bringing exciting changes with them.

One of the key areas of effort is the traditional Italian wine varietals and production. They were the first grower and producer of Nebbiolo in modern U.S. history, and have the most Nebbiolo acreage in the USA, having produced this storied Piemonte varietal here since 1982. They have been at the cutting edge of the "Cal-Ital" phenomenon; the first producer of a Super Tuscan styled wine "Etrusco" a gorgeous blend of Sangiovese and Cabernet Sauvignon, as well as a lush, many-layered Sangiovese. Nebbiolo comes into its own at Martin Weyrich, who, along with very low (less than three tons per acre) yields, utilizes the ripasso method (macerating the newly made wine on the skins and pulp of previously vinified Nebbiolo) which adds color, tannin and mid-palate weight to the final product. The final wine is unique in Paso Robles: with floral and red fruit aromatics, very spicy, with a delicious earthy complexity.

The winemaking team of Craig Reed, who has made the wines since 1994, and Alan Kinne, on board since 1999, has been a productive association from the start, each playing off the other's strengths. Reed, with a UC Davis background, was the protégé of Nick Martin (of the original Martin Brothers winery) and gave him the bug for Nebbiolo. He still attends VinItaly every year, and has toured wineries in all the major regions of Italy, in addition to his other global travels. Alan Kinne, a "Michigan farm boy" with a degree in literature and philosophy, who started his career at Tabor Hill Winery in Michigan, was a leading eastern U.S. "wine wizard" having been behind some remarkable successes in Virginia with Horton Viognier and Piemonte Reserve Chardonnay. His wide travel and studies of global enology and viticulture have proven a useful perspective for the winery. They also have an ongoing innovative and aggressive barrel program, with multiple luxury cooperages supplying the best barrels for the trials, mostly of for the red wines, which are kept separate until final blending.

This team is producing a remarkable range of wines, all in gorgeous Italian glass bottles, which they import directly. In addition to the very trendy Pinot Grigio, there's a bright, fresh unwooded Chardonnay from their estate fruit at Huerhuero Creek, and an Edna Valley Chardonnay (a barrel-fermented style) from Edna Valley vineyard fruit. They are well known for their slightly sweet, floral Moscato Allegro. On the red side, the Insieme Vini Rossi (Sangiovese, Nebbiolo, Zinfandel, Cabernet Sauvignon and Barbera) illustrates the meaning of the Italian/Piemontese dialect words "together" and "seamless." Sangiovese Il Palio and a Zinfandel La Primitiva from old, dry farmed, head pruned vines in the Santa Lucia Mountains. They also make a single vineyard Zinfandel, from the Dante Dusi Vineyard. Their two Nebbiolos, all from their estate fruit: the Il Vecchio is noticeably more extracted and spends 18 months in French oak. Rounding this out is the Etrusco Cabernet Sauvignon, blended with around 15% Sangiovese, a supple and lush approach. It would not be an Italian portfolio without the fiery but smooth spirit of Grappas, which they make from Nebbiolo and Moscato pomace. They also produce a sparkling wine!

As if all this that was not enough to get your attention, Martin and Weyrich has one of the most sought after, elegant Bed and Breakfast operations in the entire Central Coast, and a delightful, always busy tasting room as well. This remarkable story continues to grow into a world class Paso Robles destination winery.

MARTIN AND WEYRICH WINERY
P.O. Box 7003
Paso Robles, CA 93447
Phone: 805-238-2520
Fax: 805-238-6401
Web: www.martinweyrich.com
Hours: Sunday-Thursday 10 a.m.-5 p.m.;
Friday & Saturday 10 a.m.-6 p.m.
Case production: 80,000

NINER WINE ESTATES

Dick Niner is a turnaround specialist. He spent 30 years developing and motivating management teams to fulfill their company's potential, in spectrum of industries that ranged from military electronics to medical devices to school supplies, and more. Over the course of time, he investigated several vineyards and wineries, both for himself and other investors. He purchased a sunglass marketer and manufacturer in 1996, in San Luis Obispo, he and his wife Pam quickly became enamored with the Paso Robles region.

The opportunity to purchase a property he calls Bootjack Ranch, a 224-acre property on the east side in 1999, came up, and the story of Niner Wine Estates begins here. They have purchased Heart Hill, a 139-acre parcel near the western edge of the Paso Robles AVA, 12 miles from the Pacific Ocean. There is actually a heart shaped grove of oak trees on the side of one of the rolling hills on the property. They plan to source their Syrah, Malbec and Petite Verdot from here.

Bootjack Ranch had 54 acres planted to Sangiovese, Syrah, Merlot, Cabernet Sauvignon, and a hybrid called Chancellor. On the advice of well-known vineyard manager, Jim Smoot, their vineyard consultant, they grafted the Chancellor over to Barbera, Cabernet Franc and Sauvignon Blanc. They have since increased the plantings by another 61 acres, mostly to Bordeaux varietals and have added the Heart Hill vineyard to their portfolio (see photo above). Together they comprise over 360 acres, of which 125 are now planted to almost a dozen varietals, planted with strict attention to low yields and ensuring the superb quality of the final product.

In addition to the legendary Jim Smoot, Dick brought in Chuck Ortman, founder of Meridian (you would have been on the moon for the last 25 years not to hear of that winery, now owned by a multinational conglomerate) to consult on winemaking. He hired Amanda Cramer, a New Hampshire native that left a teaching post to go to UC Davis and become a winemaker. After Davis, she spent three years working with Robert Mondavi wineries, in the Napa Valley, and the wonderful D'Arenburg winery in McLaren Vale, South Australia, as well as Casa Lapostolle in Santa Cruz, Chile. She worked under Heidi Barrett at Paradigm as well, honing her skills with Bordeaux varietals, and came to work for Niner in 2004.

Brian Storrs was hired as president, coming over from very successful wine sales and marketing positions at Beringer, Brown Forman and Wild Horse, and is already having an impact, judging from the sleekly packaged, very 21st century look of the wines.

Ms. Cramer's wines all have an elegant, modern feel to them, but with that classic Paso Robles rich, full-bodied attitude. The Bootjack Ranch supplies the current releases...there is a 100% varietal Sauvignon Blanc, partially barrel and stainless steel fermented. She made a saignee of Barbera, called Rosato, in an off dry style (this sold out immediately) as well as a dry red Barbera, blended with a little Sangiovese, that spends some time in French, Hungarian and American oak barrels. There is a Sangiovese, blended with a little Merlot, and a beautiful 100% Syrah, that she makes using about 20 percent whole berries, open top fermenters with thrice daily punchdowns. It spends a year and half in one third new, mostly French (and some Hungarian) oak barrels. They have released a Cabernet Franc, with a little more new French and Hungarian oak, and a classic Bordeaux styled Cabernet Sauvignon, blended with Merlot and Cabernet Franc, aged for 16 months in 50% new French and Hungarian barrels.

Plans are afoot for a state of the art winery to be built on the Heart Hill property, with a new tasting room, a demonstration kitchen for visiting chefs, a private wine club lounge and patio facilities. Dick and Pam Niner's rapidly developing vision is scheduled for a grand opening in summer of 2008. We can't wait.

NINER WINE ESTATES
1322 Morro Street (temporary address)
San Luis Obispo, CA 94301
Phone: 805-239-2233
Fax: 805-239-0033
Web: www.ninerwine.com
Email: info@ninerwine.com
Case production: 5,000

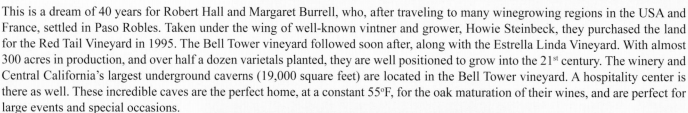

ROBERT HALL WINERY

This is a dream of 40 years for Robert Hall and Margaret Burrell, who, after traveling to many winegrowing regions in the USA and France, settled in Paso Robles. Taken under the wing of well-known vintner and grower, Howie Steinbeck, they purchased the land for the Red Tail Vineyard in 1995. The Bell Tower vineyard followed soon after, along with the Estrella Linda Vineyard. With almost 300 acres in production, and over half a dozen varietals planted, they are well positioned to grow into the 21st century. The winery and Central California's largest underground caverns (19,000 square feet) are located in the Bell Tower vineyard. A hospitality center is there as well. These incredible caves are the perfect home, at a constant 55ºF, for the oak maturation of their wines, and are perfect for large events and special occasions.

The well-known Texas winemaker Don Brady joined the Hall team in July 2001, moving here with his wife, Kasi, and their two children. A Texas Tech graduate, with more than 20 years of experience and multiple awards, including the worldwide acclaim with the International Wine and Spirits "Wine of America" award, he helped establish the Texas viticultural scene. His wines were featured at several White House functions with Ronald Reagan and George Bush; and were served to Queen Elizabeth II and Prince Charles during their visits to America. His goal is to capture the essence of the vineyard, and he has an excellent palette.

All the fruit is hand harvested in small lots with minimal handling. They utilize new state of the art equipment, as well as a traditional approach. The wines are balanced, elegant, with that streak of Paso richness running through them. This is a winery to watch.

ROBERT HALL WINERY
3443 Mill Road
Paso Robles, CA 93446
Phone: 805-239-1616
Fax: 805-239-2464
Web: www.roberthallwinery.com
Hours: Daily 10 a.m.-5 p.m.
Case production: 30,000 cases

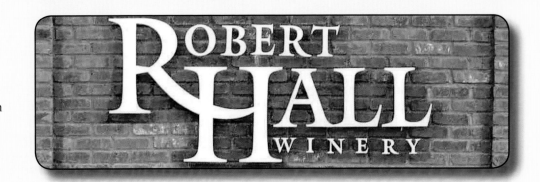

RN ESTATE VINEYARD & WINERY

Six and a half miles north of Paso Robles, at almost 1,000 foot elevation, on a west ridge of the Estrella River valley, is Roger Nicolas' 40 acres of heaven. A native of Brittany, France, he arrived in New York at the age of 20 to begin his own version of the American dream, working at a series of prestigious restaurants such as La Grenouille, and L'Etoile in San Francisco. There, he opened his first restaurant, La Potiniere, and went on to open the renowned Home Hill Country Inn, in New Hampshire, now owned by the Relais and Chateaux group.

"Retiring" at the age of 50, he sought to indulge his passion for wine, and began a search for the perfect land for a vineyard. He purchased 40 acres in Paso Robles, and planted vines on four, primarily to the main Bordeaux varietals, Cabernet Sauvignon, Cabernet Franc and Merlot, with a little Syrah, Mourvedre, and Zinfandel. He also grows a tiny bit of Viognier and Roussanne, currently at experimental levels. His property is sustainably farmed, mostly by Roger himself with a hoe in hand.

Most of his production is estate grown, currently produced at Paso Robles Wine Services. He is self-taught, though a veteran of many European Union (EU) and UC Davis winemaking courses and seminars. He utilizes a stainless steel basket press, open top fermenters, and quite a bit of new and neutral French oak, with some American and Hungarian barrels as well.

He explained in his soft French accent that he "...likes to make his wines like sauces, not gravies...," eschewing the chewy, overripe qualities that tend to pervade current California wine styles. His Zinfandel is a unique, food friendly wine, and he is particularly proud of his current red blends: Cuvee de Trois Cepages, a Cabernet Sauvignon, Cabernet Franc, Merlot blend, and westside Cuvee, again mostly Cabernet(s) Sauvignon and Franc, but with some Merlot, Petite Verdot and Malbec. Both spend about 20 months in French oak.

He makes a Cuvee des Artistes, based on Syrah, with Cabernets Sauvignon and Franc, blended with a little Zinfandel...utterly delicious after 20 months in American, Hungarian and French oak. There is a Syrah/Mourvedre blend, and "Young Vine" varietal Mourvedre and Zinfandels. He generally gives the wines a 2-3 day cold soak, but this is no hard and fast rule. All his wines are unfined and unfiltered. He also has released a Santa Rita Hills Pinot Noir, from fruit sourced from the Fiddlestix Vineyard.

I got the distinct impression that Roger is a private sort of fellow, relishing his current bucolic country lifestyle after 30 years in the limelight, dealing constantly with the public. That said, he is more than willing to make appointments for tasting and tours of his lovely property, but given the small lot production (current production is 800 cases total, 400 from his own fruit) he will no doubt be selling out his wines from vintage to vintage.

RN ESTATE VINEYARD & WINERY
7986 North River Road
Paso Robles, CA 93446
Phone: 805-467-3106
Web: www.rnestate.com
Email: rnicolas@rnestate.com
Hours: By appointment only
Case production: 800

CHAPTER NINE: GENESEO DISTRICT

GENESEO DISTRICT (approximately 17,300 acres): warmer region III-IV transitional climate, with marine incursion through the Templeton Gap area during part of the growing season, but with summer daily minimum (night time) temperatures warmer, with mixing by the winds and cold air drainage down the hillsides, highest and oldest terraces of the Estrella River and Salinas/Estrella/Huerhuero river confluence, with uplifted Huerhuero Hills pushing through these terraces through time, with geology Tertiary to Quaternary Paso Robles Formation, older river deposits listed above, and more recent alluvial deposits of Huerhuero Creek, old clay loam to silty clay loam Alfisols and Mollisols, with buried cemented horizons, oak savannah vegetation (Elliott-Fisk, 2007).

CASS WINERY

A freshly minted Paso Robles story, located in the Geneseo District, a little off the beaten track...Cass Winery is making waves as one of the newest 100% estate bottled, Rhone specialist wineries in Paso Robles. Unique in more ways than one; from the striking, modern southwest architecture of the winery and tasting room, to the ENTAV clones of Rhone varietals. ENTAV is a French Government Agency, in charge of clonal certification, that uses an extremely rigorous, sometimes decade's long clonal selection and propagation system. Cass is one of only a couple of wineries in Paso Robles that are using these precious true clones of Syrah, Mourvedre, Grenache, Viognier, Roussanne, Marsanne, and Petite Sirah.

Steve Cass is a retired financial services advisor, moving south from Walnut Creek in northern California. He and winery partner Ted Plemons have carved out 160 acres planted to Rhone varietals and built a fantastic winery to show them off. The winery and tasting room, under one roof, is at once striking and welcoming. The is a wine lined library that seats 16, a larger barrel room, fully ensconced next to a brand new commercial kitchen. World-class picnics can be taken outside under a huge oak and wisteria trellis. All are laid out with the synergy of wine and food in mind, and they have already hosted a number of cooking classes.

The original winemaker, Dan Kleck, who crafted the 2003 and 2004 vintages, has hired Lood Kotze, a native of South Africa. His first releases are already garnering awards, specifically a gold medal from the largest wine competition in the USA, the Orange County Fair for their amazing 2005 Viognier.

These wines are bold, varietally true and so expressive of classic Paso Robles *terroir*. They blended 20% Syrah into their Cabernet, and after harvesting a minuscule 100 pounds per acre, crafted an inky, peppery, classic Northern Rhone Style Syrah. A number of blends round out the red picture "...Rockin' One" (no typo here) a traditional Southern Rhone style blend of Grenache, Syrah and Mourvedre, and Hacienda, from a vineyard directly in front of the winery...a blend of Mourvedre and Grenache. They also produce a gorgeous Roussanne. For rose fans, there is CassaNova Rose Cuvee, again made from Mourvedre, which always sells out.

The wines are hand made in small lots, hand punched, and the oak influence is reigned in. In a word, Beautiful. Steve Cass, Ted Plemons, Dan Kleck and Lood Kotze have everyone's attention now. They will cap production at 2,000 cases, and are looking forward to a future where all of their wines will be sold out of the winery, which will no doubt be soon. Bravo to them all!

CASS WINERY
7350 Linne Road
Paso Robles, CA 93446
Phone: 805-239-1730
Fax: 805-227-2889
Web: www.casswines.com
Email: info@casswines.com
Hours: Monday-Friday noon-5 p.m.,
Saturday-Sunday: 11 a.m.-6 p.m.
Case Production: 2,000

CLAUTIERE VINEYARD

A unique scenario: "Edward Scissorhands meets the Mad Hatter at Moulin Rouge." This is what is waiting for you at the Clautiere tasting room, created by Claudine Blackwell and Terry Brady after moving to Paso in 1999. Their eclectic backgrounds of restaurant owners (The Lobster on Santa Monica Pier...if you have not been there, it is a must), landscape designer, welder, fashion designer led ultimately to the current incarnation of winegrower and winemaker.

They created a fanciful world out of an old local ranch, bursting with color and vitality. The old farmhouse is now an amazing tasting room, gift shop and commercial kitchen. There is also a theater space for special events. The exterior boasts 230 feet of metal fencing and custom metal sculptures that Claudine designed and welded, adding a final shine to the whole picture.

Claudine is the vineyard manager, overseeing an eclectic selection of Bordeaux, Rhone and Portuguese varietals. They produce some unique wines for Paso Robles: a Port from Touriga Nacional, Tinta Cao and Tinta Roriz, from a one-acre parcel on the property. There is also an Estate Grenache, and three interesting red blends; Grand Rouge, a Syrah/Cabernet/Rhone varietal blend, Mon Beau Rouge, a more classically styled Southern Rhone style blend, and Mon Rouge, a Syrah, Cabernet and Mourvedre blend. They also produce a gorgeous Viognier and a Cabernet Sauvignon as well.

In 2005, the first crops of Roussanne, a classic white Rhone varietal, were planted along with Cabernet Franc, Petite Verdot, and Malbec for an upcoming Bordeaux blend. This is a fascinating new chapter in the Paso Robles story, sure to evolve in a most interesting future of excitement and merriment.

CLAUTIERE VINEYARD
1340 Penman Springs Road
Paso Robles, CA 93446
Phone: 805-237-3789
Fax: 805-237-1730
Web: www.clautiere.com
Email: info@clautiere.com
Hours: Daily noon-5 p.m.
Case production: 2,700

CHUMEIA VINEYARDS

Lee Nesbitt really wanted to be a middle linebacker for the Minnesota Vikings. He attended college at Cal Poly SLO on a football scholarship, actively pursuing his goal. Eventually he realized that maybe he needed to rethink his strategy, and spent a decade patiently researching and planning the next phase of his life. He interned at J. Lohr, and crafted his first wines under his own Dry Canyon label. After selling the Dry Canyon wines, he worked at the Meridian winery retail operation to garner all the knowledge he could about marketing, later transferring to a wine technician position, to hone the enology skills he would acquire there. In 1999, several investors later, he collaborated with Mark Nesbitt and John Simpson. Drawing on Simpson's vast viticultural experience and his

father Mark Nesbitt's agribusiness acumen, they have positioned themselves to utilize Paso Roble's potential. The initial purchase was 20 acres of East Paso Robles land, with an additional 11 acres planned. They also have a long-term contract with the Simpson's for their superb fruit from (unusually) low yielding vineyards in the Madera region.

The name Chumeia (pronounced "koo-may-a") is archaic Greek for "alchemy," the process of making something truly precious out of something merely valuable. By applying the four basic elements of fire, air, earth and water, and adding some well thought-out philosophy, the ancient alchemists believed they could turn essential elements, such as lead or mercury, into gold. As winemaker, Lee makes an immediate, passionate connection between the two processes. They have planted Zinfandel and Petite Sirah at their new estate vineyard, and have just released their first vintage of Cabernet Franc. They also plan to market develop strategic relationships with wineries in Argentina and Chile to increase their global exposure. Chumeia is poised for an interesting second decade.

CHUMEIA VINEYARDS
8331 Highway 46 East
Paso Robles, CA 93446
Phone: 805-226-0102
Fax: 805-226-0104
Web: www.chumeiavineyards.com
Email: lnesbitt@chumeiavineyards.com
Hours: Daily 10 a.m.-5 p.m.
Case production: 7,500

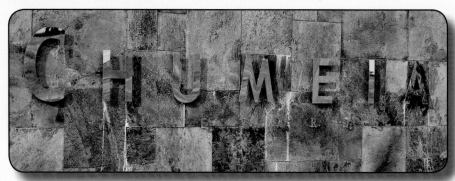

GELFAND VINEYARDS

Len Gelfand was a little testy the day I talked to him. I have to admit, calling on a winery around crush can be hazardous at best, especially if you want to talk to the winemaker. Looking at the remarkable results of his efforts, you cannot blame him. I asked him what was the one thing he wanted to get across to people, about his wines and winery, and he replied with one word "Passion." The intensity in his voice was tangible.

The Gelfand's originally hailed from southern California; Len was an insurance executive...Jan and Len Gelfand are world travelers and have loved and visited a plethora of wineries around the world. They had always dreamed of becoming the people on the other side of the counter, pouring their own wines. That dream came to fruition in 2000, upon the purchase of a 25-acre parcel in westside Paso Robles, adjacent to Penman Springs Road.

They planted the first 10 acres with Cabernet Sauvignon, Zinfandel, Syrah, and Petite Sirah, in equal proportions, very closely spaced. The resulting fruit is fantastic, and given that Len is a first time winemaker, results from his first releases were amazing. Gold and Sweepstake Gold medals from L.A., Orange County and San Diego Wine competitions, as well as a host of Silver and Bronze medals. All this from a first release! Their wines are, for the most part, almost completely sold out. All their wines are estate bottled.

Len uses all new French, American and Hungarian oak barrels, producing a Cabernet Sauvignon, Zinfandel, Syrah and a phenomenal Petite Sirah. They make a couple of red blends; Cabyrah (Cabernet/Syrah), Ménage a Bunch, also a Cab/Syrah blend, and another red blend from their best barrels, Quixotic, which they refer to as their flagship wine. SFR, another red blend, will vary in composition from year to year, is Petite Sirah-based.

They produce Sophie, named after their granddaughter who was born the day they bottled the wine. This is a late harvest Cabernet Sauvignon, fermentation arrested at about seven percent residual sugar, and sells out quickly. There is also Port style wine, also, Cabernet based, but blended with Petite Sirah and Zinfandel, the fermentation stopped with brandy made from Merlot grapes.

Gelfand is a fantastic debut from an obviously passionate new winemaker in Paso Robles. Visit the winery by appointment only to seek out these wonderful new wines from the east side of Paso Robles.

GELFAND VINEYARDS
5530 Dresser Ranch Road
Paso Robles, CA 93446
Phone: 805-239-5808
Fax: 805-239-1507
Web: www.gelfandvineyards.com
Email: len@gelfandvineyards.com
Hours: By appointment only
Case production: 1,200

MALOY O'NEILL VINEYARDS

As we've said, this book will be outdated the moment it goes to print, and this is well illustrated by the Maloy O'Neill story...They have been around for more than two decades, selling grapes to wineries in Paso Robles, Sonoma, Napa Valley, Mendocino and Lake County. It started as a 20-acre experiment in 1982, and has grown to almost 300 acres. Cabernet Sauvignon remains the flagship, but they also grow Merlot, Syrah, Zinfandel, Petite Sirah, Sangiovese, Chardonnay, Muscat Canelli, Malvasia Bianca, Pinot Grigio, and Tempranillo.

In 1998, Shannon O'Neill planted a 5-acre vineyard dedicated to varietal, clonal, rootstock and small lot experimentation, on what some said was an unplantable hill. In 2001, Shannon and his wife purchased another 80 acres, planting 25 acres of Cabernet Sauvignon, Syrah, Pinot Noir, Zinfandel and Cabernet Franc. Their total holdings add up to about close to 300 acres, spanning a wide variety of soils and mesoclimates.

President and Winemaker Shannon O'Neill prides himself on maintaining the family owned and operated business, dedicated to the quality that they've become widely known for. In addition to chemistry lab experience with the likes of Atlantic Richfield and Chevron; he is a UC Davis fermentation science graduate as well. He's running the family business full time now, having made his first vintage in 1984. He specializes in "...intense, big, extracted wines..." and consults for a who's who list of Paso Robles wineries. He makes 25-30 different wines a year under his own label, and 95 percent of the production is estate bottled.

According to Shannon, "I never met an oak barrel I didn't love..." he uses only French oak for his Pinot Noir, and a selection of American and Hungarian oak for his broad palette of wines. They specialize in small lots (<200 cases), and are scheduled to open a new winery facility in late 2006. We cannot wait to see and taste the new results from this extremely energetic family on the Back Roads of east Paso Robles.

MALOY O'NEILL VINEYARDS
5725 Union Road
Paso Robles, CA 93446
Phone: 805-238-7302
Fax: 805-226-8412
Web: www.maloyoneill.com
Email: winery@maloyoneill.com
Hours: Fri.-Sat. 10 a.m.-5 p.m.
　　　　Sun. 12-5 p.m.

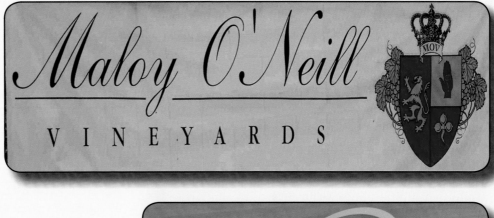

MIDLIFE CRISIS WINERY

It is an unusual, albeit honest, down to earth name for this tiny, family owned Paso Robles winery. Kevin and Jill Mittan are the dynamic duo behind everything here, from winemaking, and cellar work to marketing and vineyard management. They still commute from their jobs in the production end of the entertainment industry to Paso Robles every weekend.

They studied at UC Davis, and with a commercial winemaker in Camarillo, and spent time with winemakers and viticulturists all over the Central Coast. They had been making wine in their garage in Los Angeles for years. Their garagiste efforts won over 60 regional and international awards.

With a dual income, no kids' scenario in place, they unselfconsciously dropped everything, sold the house in Los Angeles at the peak of the real estate boom in 2004, and bought 22 acres on Highway 46 East. They have plans to

WINE (SILLY) MEDLEY

Wine spodey-odey, drinking wine;
Wine, wine all the time.

If the river was wine and I was a diving duck,
I'd dive to the bottom and drink my way back up.

Red, red wine, always on my baby's mind.
Dynamo humm, she said you can't make me drink Beer.

Just get me wasted on wine and I'm halfway there,
'Cuz when my mind's tore up, my body don't care.

Home, home on the vineyard, where the
Reds and the white wines are grown.

On Wisconsin, drink wine for your fame; and
Eat some cheese & crackers to help win the game.

If you want to be a Badger, just come along with me,
By the bright shiny light, by the blush of the wine.

I know I've missed several top forty hits and
Some obscure wine melodies, do you know others?

(Photos 35-36: New vineyard planting on Vineyard Drive, above, June 2000; and Morning fog at Halter Ranch, 22 Aug. 2002)

build an onsite tasting room and plant vineyards within the next 2-5 years. Although the water table is less than 200 feet below the property, they will be dry farming the Bordeaux varietals, Tempranillo and Zinfandel that they will be planting.

They currently make their wines, from 100% Paso Robles fruit, at Paso Robles Wine Services, under the watchful eyes of Paul Ayres of San Marcos Creek Winery. Kevin is not a fan of new oak, and utilizes as little blending as possible.

They make Pinot Grigio, Chardonnay, Syrah, Zinfandel, Sangiovese, Barbera, Merlot, and a red blend named Roo Boy after their Australian Shepard dog. Roo Boy is Petite Sirah, Merlot and Tempranillo. A new release in 2006 is Nebbiolo. They also make a couple of unpretentious sparkling wines. This kind of determination speaks to the heart of the individuality of Paso Robles.

MIDLIFE CRISIS WINERY (TASTING ROOM ADDRESS)
1244 Pine Street
Paso Robles, CA 93446
Phone: 805-237-8730
Fax: 805-237-0109
Web: www.midlifecrisiswinery.com
Email: jill@midlifecrisiswinery
Hours: Daily 11 a.m.- 7 p.m.;
or by appointment (please call)
Case production: 600

PENMAN SPRINGS VINEYARD

Another exciting story on the east side, this family-owned venture has undergone transformations in the past 20 years. The hilltop, 40-acre site was first planted in the early 1980's as Baron Vineyards, a joint venture with Kolb Vineyards, across from Union Road.

The Penman Springs that we know today started in 1996, when Carl and Beth McCasland decided to make viticulture their lifestyle, moving the whole family to Paso Robles. Upon arrival, Carl began a serious study of the viticultural potential of the land, consulting with experts, beginning a soil enrichment and vine replanting program, drilling new water wells, and refurbishing the existing vine trellis systems.

Carl and Beth have expanded the vineyards, planting more red varieties, adding a five-acre block of Syrah, and grafting over another five acres to Petite Sirah, which makes a fruit-exploding wine, a favorite of many local wine lovers. Paying particular attention to trellising and varietal combinations makes a big difference in the final wines. Carl utilizes several different combinations of trellises with the varietals, maximizing sun and fruit exposure, managing leaf canopies, and insuring the individual characteristics of the fruit are fully expressed.

Carl acquired winemaking equipment in the past decade, and hired winemaker Larry Roberts. The first crush was in 1998, with their own Merlot and Cabernet Sauvignon. They have since released fortified Muscat in the Aussie style as well. The 1999 Muscat took a silver medal at the San Francisco International Wine Competition in 2004. The potential for Petite Sirah was soon known to all that tasted it, and the 2001 Estate Petite took a bronze at the same competition. The Penman Springs tasting room opened in the fall of 2000 in a beautiful home-like atmosphere. They produce about 500 cases of artisanal quality, classic Paso varietals.

Rosie and Beth, and other assorted friends, manage their lively tasting room, which is always full of good humor and stocked with delicious cheeses and original recipe treats to complement the wines. A visit to their tasting room is a must for anyone who enjoys good company, great wines and a beautiful view of Paso.

PENMAN SPRINGS VINEYARD
1985 Penman Springs Road
Paso Robles, CA 93446
Phone: 805-237-8960
Fax: 805-237-8975
Web: www.penmansprings.com
Email: carlmc@tcsn.net
Hours: Friday-Sunday 11 a.m.-5 p.m.
(Closed January)
Case production: 500

STEINBECK VINEYARDS

If you want to explore the roots of wine in Paso Robles, a visit with Cindy Newkirk at the Steinbeck Vineyard WineYard is imperative. This historic Paso Robles family goes back for five generations in the area, first arriving in 1884 from Geneseo, Illinois. The Ernst family (her maternal grandmother, Barbara Ernst's side of the family) was the focus in the nineteenth century wine industry of Paso Robles. They were well known for the superb quality of their grapes and wine even then. The Ernst family branch originally hailed from Alsace in northeastern France on the German border.

Like so many other immigrants, they said the difficult goodbyes to ancestral lands, homes and families, setting off for America in 1868. Seventeen days on a steamer across the Atlantic to New York, and another tough five days to get to Chicago, they finally landed in Geneseo, Illinois, where Barbara met and married William Ernst. They heeded the call to go west after seeing a newspaper article asking families to start a Lutheran Church on "good farm land". In 1884, they packed again and moved west. The Ernst brothers were named after their hometown, and lived in Geneseo, California for generations.

Son Frank and his wife Rosetta Ernst built the ranch house in 1923, raising five children. It passed to daughter Hazel and husband George Steinbeck, and then onto Howie and Bev Steinbeck; all raised families here. Tim and Cindy Newkirk, Howie's daughter, and their children live in this house today, extending this remarkable five-generation legacy.

Howie Steinbeck had farmed barley on the original 50-acre parcel with his grandfather Frank, since 1943. Howie and wife Bev (nee Jespersen) bought the property in 1972, planting vines in 1982. They specialize in Cabernet Sauvignon, Petite Sirah and Viognier. They now farm 500 acres of premium vineyards, both selling fruit to a long list of luxury fine wineries all over California. In total, they farm or manage over 1,000 acres of vineyards. They take stewardship very seriously, practicing sustainability, strict canopy and high tech irrigation management, balanced vine structure, to keep yields low and quality high. "We want it safe for our families and employees..." Cindy explains, "...my goal is that my children's children will be raised here..."

They grow thirteen varietals including Roussanne, Muscat Canelli, Sauvignon Blanc, Zinfandel and Syrah. They also grow the classic Portuguese varietals, Tinta Madera and Touriga. Eberle makes a varietal and a port style wine from this fruit. They have 100 acres of own rooted Cabernet Sauvignon and Chardonnay. These vineyards have supplied vegetative cuttings to many vineyards around Paso Robles.

The WineYard at Steinbeck Vineyards is a fantastic concept...Cindy has developed a multi-layered presentation, available by appointment only, broken out into four parts: the family history in the region, the overall picture of Paso Robles and how it fits into the California wine industry, then the common sense practices that go into premium wine grape vineyard management, and an exploration of the growing season. She includes an extensive Jeep tour of the vineyards, as well.

They have broken ground for their winery facility, and will be able to make and bottle their own wines...they have slated several different vineyard Cabernet Sauvignon blocks, a Petite Sirah, Viognier, Zinfandel and Merlot. Steve Glossner has been tapped to make the wines. He will use some new Hungarian oak barrels for the red wines, and stainless steel for the Viognier.

The new tasting room will reflect their history. "It will be as though you are walking into our ranch home...a 1920's ranch home...I want people to feel as if they are walking into our past...and I hope the wine is as good as the concept..." Their first releases will total under 1,000 cases, and then increasing to a production of about 7,000 cases.

"My dad had tremendous vision twenty five years ago...back when people just thought he was crazy...and I'm seeing what will take us to the next level. My son, who is in his twenties, is working with us full time...there are three generations working successfully

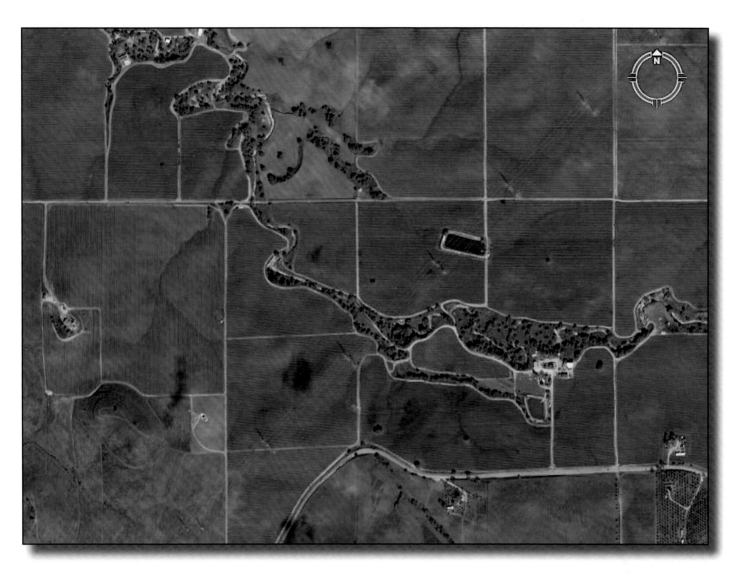

Google Earth image of Steinbeck Vineyards and the surrounding lands north of Union Road in the Geneseo District.

Photo 37: Cindy Newkirk and her father, Howie Steinbeck, on their vineyards north of Union Road (10 November 2006).

together now. I'm especially proud of that...," Cindy explains. This story is the real thing, and new chapters are just beginning for the Steinbeck Vineyards and WineYard. It is truly a classic Paso Robles pioneering story.

THE WINEYARD AT STEINBECK VINEYARDS
(NO drop-ins, please...by appointment only)
5940 Union Road
Paso Robles, CA 93446
Phone: 805-238-1854; 805-674-1909 (Cindy's cell)
Fax: 805-238-7327
Web: www.thewineyard.com
Email: thewineyard@starband.net

TOBIN JAMES CELLARS

At the eastern end of the Estrella Terraces, is a textbook example of Paso style and energy. Tobin ("Toby") James, a Cincinnati raised, and California native who decided to make his own way after years at Estrella, with Gary Eberle, then with Gary at Eberle Winery, finally as the founding winemaker at Peachy Canyon in 1987, a local producer came to Toby with several tons of Zinfandel grapes that were refused by another winery. He accepted the fruit, made the wine, and won many medals with it.

In 1989, Peachy Canyon agreed to let him use their winery to make his own wine, in exchange for his winemaking efforts on their behalf. In 1994, Toby purchased 41 acres at the end of Union Road, and Paso Robles has never been the same. The winery was built from the ground up on the site of an old stagecoach shop, delightfully and historically festooned western style, and is home to a magnificent 1860's Brunswick mahogany bar from Blue Eye, Missouri. They are trying, and succeeding, given their 14,000+ member "James Gang" Wine Club, in creating a classic western, fun and hospitality-oriented winery, with some fine wines to boot. By the way, the wine club benefits are not to be missed; great wines at discounted prices and a gift in every shipment. I love the coffee mugs from the last shipment...just for coffee in the morning, but who knows what vintage may appear in them each night!

James' business partners since 1997, Lance and Claire Silver's business and marketing and practical understanding have dramatically increased the brand presence, on wine lists and fine wine retail shelves. They make close to 20 different varietals; the ubiquitous Chardonnay (sourced from Monterey fruit), and "Made in the Shade" Merlot (Paso Robles AVA), dynamite "Ballistic" and numerous other Zinfandels sourced from some of the best vineyards in Paso Robles (including the famed Dusi Ranch). The disarmingly named "Chateau le Cacheflo" is an interesting, delicious blend of Syrah, Dolcetto and Refosco, all from Paso Robles grapes. Joined in this lineup are a "Rock and Roll" Syrah, James Gang Reserve Zinfandel and Cabernet Sauvignon, and a "Liquid Love" Late Harvest Reserve Zinfandel. There is also "Charisma", another extremely popular dessert style Zin, albeit in a lighter style, with a gorgeous package.

They are deservedly well known for their ongoing series of winery events, and a visit to Tobin James during those Paso Robles wine festivals is a must. Just remember to reserve tickets early for their events, since they are often "Sold Out" shortly after the event advertising notice goes out! You may very well run into Toby, Lance or Claire and be corralled into a barrel room tour or, at the very least, to buy some of these delicious wines and some other fun gifts. Participation in their winery events will be time well spent and great fun-filled experience for wine lovers and social butterflies.

TOBIN JAMES CELLARS
8950 Union Road
Paso Robles, CA 93446
Phone (805) 239-2204
Fax (805) 239-4471
Web: www.tobinjames.com
Case production: 30,000

VINA ROBLES

In 1996, yet another European incursion made its way to Paso, when Hans Nef and Matthias Gubler, natives of Switzerland, came to California, and quickly discovered the burgeoning potential of Paso Robles. The first purchases of three estate vineyard properties started them off; the Pleasant Valley, Jardine and Huerhuero vineyards were all soil mapped and planted with over a dozen different varietals, the majority being Cabernet Sauvignon and Syrah. Phil Christensen of Agriglobe, a Fresno based farm management and consulting firm, guided the vineyards through the initial phases. Agriglobe is owned by Christensen and Hans Michel, who is also Nef's business partner, long time friend, and compatriot.

They decided early on to hand prune and hand harvest the fruit from these vineyards, but with an efficient, modern (Swiss!) turn; some next level vineyard practices such as the use of barcodes to track data from each vine row, as well as the use of state of the art soil moisture sensors to monitor irrigation. All of this means better fruit, which in turn means better wine.

Matthias Gubler joined the Vina Robles team in 1999, bringing extensive European winegrowing and winemaking experience. From his start on his family's Pinot Noir vineyard in Switzerland, he honed his skills in California, the Rhone, and Tuscany wineries. He is very involved with all aspects of vineyard management, as well, and prefers to be called a "winegrower" rather than "winemaker."

These wines represent "new Paso" style, with ripeness and balanced tannin. They want to produce wines that are delicious on release, and will continue to evolve in the cellar. There is an Estate Sauvignon Blanc, Cabernet Sauvignon, and Syrah, which are made from a blend of fruit from three vineyards. They also produce single vineyard wines from the Jardine and Huerhuero vineyards: Petite Sirah and Cabernet from Jardine, and a Syrah from Huerhuero. They have also planted and produce Tempranillo, Tannat and Touriga, all red grapes from warm European climates, well suited to the warmth of eastern Paso Robles. Here's the extensive list of varietals; Cabernet Sauvignon, Cabernet Franc, Chardonnay, Grenache, Malbec, Merlot, Mourvedre, Petite Sirah, Sauvignon Blanc, Syrah, Tannat, Tempranillo, Touriga, Viognier, and Zinfandel.
The hospitality center opened in July 2007.

VINA ROBLES
P.O. Box 699
Paso Robles, CA 93447
Phone: 805-227-4812
Fax: 805-227-4816
Web: www.vinarobles.com
Email: m.laderriere@vinarobles.com
Case production: 15,000

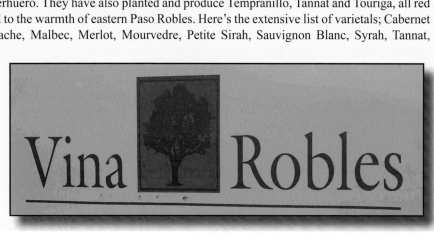

"WINE THAT MAKETH GLAD THE HEART OF MAN."
THE BOOK OF PSALMS, 104:15

EL POMAR DISTRICT (approximately 21,300 acres): moderate region II climate with airflow through and across Templeton Gap area part of the growing season, higher and older terraces east of the Salinas River, with geology Miocene Monterey Formation to Quaternary Paso Robles Formation and more recent alluvial deposits of Huerhuero Creek and Salinas River, grassland Mollisols with well developed surface horizons, with shallow bedrock in some areas covered by alluvium, oak savannah vegetation (Elliott-Fisk, 2007).

FRALICH VINEYARD & WINERY

In 1992, Rhone varietals were not on the radar screen for too many wineries in Paso Robles. But that's what Harry Fralich planted on his 20-acre property in Templeton, originally purchased in 1980. Today they (and now many other wineries) grow a roster of the usual Rhone suspects; Syrah, Viognier, Roussanne, Marsanne, and have added Petite Verdot and Verdelho, the white grape of Portuguese extraction, originally (and still) grown as a key component of Madeira. Yes, Dorothy, there is Zinfandel too.

Harry and his wife, Ruth, are refugees from LA, where Harry worked for Hughes aircraft until 1993. It took a few years of the five-hour drive back and forth between the vineyard he established in 1989, and LA, to realize that it was time to make a move. That's exactly what they did, and haven't looked back.

The fruit is grown with great care on their estate, all sustainably farmed, with particular attention paid to low yields...utilizing hand pruning, leave and shoot thinning and hand harvesting. He gets about 3-4 tons per acre, and all but 10 percent of his fruit is snapped up by a laundry list of who's who Rhone varietal producers, many of whom bottle with a Fralich Vineyard designate on their labels.

In 2003, he decided to take that 20 percent of his own production and make some wine. This is always going to be a showcase for his own version of Paso Robles *terroir*, and always an extremely limited number of cases, less than a thousand. Harry's been making wine since 2002, but his current consulting winemaker is Steve Glossner, well known for some phenomenal releases at Adelaida Cellars, Justin and Halter Ranch; at printing all of the Fralich releases are sold out except for the Syrah... They also produce a Viognier, "Harry's White Blend" of Viognier, Roussanne and Verdelho, "Harry's Patio Wine" mostly Viognier and a dash of Verdelho, and "Harry's Patio Red" a Zinfandel/Syrah/Cabernet blend. Rounding out this mouth-watering portfolio is a Claret (deep rose style) of Syrah, and finally Harry's Tears, a late harvest Verdelho.

All I've got to say is...he started taking gold medals the moment these wines came on the competition scene, selling out rapidly, and production isn't going to get any bigger. They'll be bottling an "expensive and out of season" Shiraz, a port style wine, and more whimsically named cuvees are in the pipeline. Harry's taken a great location, made it his own, focused persistently on quality and expression of *terroir* in the vineyard, and the reaction has been swift and sure. Try to obtain some of these wines. If you want to visit (sometime soon) give them at least two weeks notice to make an appointment. What a nice problem for Fralich to have!

FRALICH VINEYARD & WINERY
P.O. Box 818
Templeton, CA 93465
Phone: 805-434-1526
Fax: 805-434-3379
Web: www.fralichvineyard.com
Email: harry@fralichvineyard.com
Hours: by appointment only

McCLEAN VINEYARDS

Mike and Judy McClean met at, of all places, a wine tasting class at Cal Poly SLO in 1978. As their love for red wine and each other grew, an opportunity came up to purchase a very distinctive plot of land along the South El Pomar corridor. There was an old almond orchard growing on it, completely overgrown...they recognized the potential immediately, and in 1994 purchased this 18-acre parcel.

After clearing the land, they planted old Australian clones of Syrah on their own roots, which did wildly well. So well, in fact, that Ken Volk, who was still at Wild Horse, came to visit one day, and after a day in the vineyard, offered them their first contract for grapes. This phenomenal fruit became the backbone of Wild Horse Syrah for several years. They still hand tend the entire parcel themselves, each vine receiving personal attention to the smallest detail of pruning, trellising, and harvest.

With success right out of the box, it wasn't long before they decided to produce their own wine, a single vineyard, one-of-a-kind reserve Syrah. Mike uses about a 60/40 split between medium toast French and American oak, rotated regularly, about 15% new

every year. The wine undergoes a month long maceration, (no cold soak necessary with this fruit...) after 20-23 months in oak, the wine is lightly filtered and bottled. They awards rolled in immediately, and this year they released a saignee of Syrah, a gorgeous, bone dry Rose style that spends ten months in neutral oak barrels.

This is the kind of single-minded focus that will drive the reputation of Paso Robles into the 21st century. Make an appointment to meet this charming couple and taste their amazing Syrah.

McCLEAN VINEYARDS
4491 S. El Pomar Drive
Templeton, CA 93465
Phone: 805-237-2441
Fax: 805-237-0179
Web: www.mccleanvineyards.com
Email: mccleanwine@netzero.net
Hours: By Appointment Only
Case production: 3,500

VICTOR HUGO WINERY

No literary connection here. This is a great little operation run by Vic and Leslie Roberts, off the beaten track on El Pomar Road near Templeton. Vic's dream began to take shape in 1985 after an enology degree at UC Davis, and 15 years at another Paso winery. He and Leslie planted 15 acres on the (now) home property. A working windmill is highly visible and they have renovated a 100-year-old barn as well. Their holdings have expanded to close to 80 acres.

They planted Chardonnay, Zinfandel, Syrah, Petite Sirah, Viognier, and the classic Bordeaux 5; Cabernet Sauvignon, Cabernet Franc, Petite Verdot, Merlot and Malbec. They carefully tend vines on the shaly hillside soils with particular attention to low yields, utilizing vertical shoot positioning (VSP) trellises, leaf pulling, shoot and cluster thinning.

All the vines are grafted onto drought tolerant, and nematode- and phylloxera-resistant rootstock. They make use of mild deficit irrigation to concentrate flavors in the fruit as well. Their Templeton Hills vineyard is the source for their estate bottlings; 100% varietal Merlot, Zinfandel, and a Syrah with a whiff of Viognier (a'la the Northern Rhone style). There is also a pure Petite Sirah, showing all the promise this varietal does in Paso Robles. In addition, they also purchase fruit to add to their own estate Viognier.

All these wines are hand harvested and sorted, small lot fermented, and in addition to punching down, go through extended maceration and long aging (15-22 months) in French and Hungarian oak barrels, before final blending and bottling.

They also make a couple of very interesting Bordeaux style blends; " The Hunchback" which is mostly Cabernet Franc and Malbec, with the remainder made up of Petit Verdot. There is also "Opulence" made from all five classic Bordeaux varietals. These wines spend up to two years in oak, and are very limited. They are excellent examples of their dedication to producing rich, elegant, long-lived wines. Judging from the lengthy awards list from a who's who of national wine competitions, they have the secret to success.

VICTOR HUGO WINERY
2850 El Pomar Drive
Templeton, CA 93465
Phone: 805-434-1128
Fax: 805-434-1124
Web: www.victorhugowinery.com
Email: sales@victohugowinery.com
Hours: By appointment only-please call
Case production: 2,500

"FROM WINE WHAT SUDDEN FRIENDSHIP SPRINGS!"
JOHN GAY

CRESTON DISTRICT (approximately 47,000 acres): low region II-III climate, warmer than El Pomar, with most modest maritime influence but pronounced cold air drainage downslope, alluvial terraces and benches of Huerhuero Creek and Dry Creek at northern base of La Panza Range, with geology Tertiary to Quaternary Paso Robles Formation and alluvial deposits of diverse ages, diversity of alluvial and some bedrock soils exposed with erosion, natural vegetation open to more closed oak woodlands to chaparral on highest slopes (Elliott-Fisk, 2007).

CHATEAU MARGENE

East of Paso Robles, near Creston, is a very new, unique, small property. To use their own phrase, a micro-property. Chateau Margene, owned by Michael and Margene Mooney, this six-acre parcel is true to its Bordeaux theme name, specializing in Bordeaux blends of Cabernets Sauvignon and Franc, Merlot, Petite Verdot and Malbec.

Michael is the winemaker, a great believer in low yields, deficit irrigation and strict canopy management to produce superb fruit. They also shoot and cluster thin, "...great wine starts with great fruit...it all begins in the vineyard...," Michael says. They also have close relationships with some great vineyards, from which they source fruit, located in various mesoclimates within the Paso Robles AVA.

These luxury, small lot wines are very much reflective of their artisanal approach; the Reserve Cabernet is 85% Cabernet Sauvignon, the remainder Cabernet Franc and Merlot...cold soaked, indigenous yeast, hand punched in small open top bins, then sur lies for 15 months. This is followed by 12 months in partially new French oak. It is unfined and unfiltered. The Cabernet Sauvignon is sourced from their own grapes in the Casa Grande vineyard, with a dash of Cabernet Franc from the Smoot Vineyard. It spends 15 months resting sur lies, and 13 months in oak. Michael produces a Pinot Noir, sourced from the tiny, cool westside vineyard near York Mountain; 100% Pinot Noir, aged in mostly neutral French oak, and sells out rapidly. There is a Syrah/Cabernet Blend, and a Chardonnay to round out the portfolio. All his wines are round, supple and complex, with a restrained touch of oak...

Beau Mélange is their flagship wine. This estate fruit only, Bordeaux varietal red blend spends 16 months sur lie, and then is racked off to more (50% new) French oak about 16 months...it is expensive and delicious, sold on a pre-release and allocated basis to wine club members only.

This is the kind of care and concern that only the small, family owned wineries can accomplish. They will increase production to 3,000 cases, and Michael's sons, Chris and Jon, have also started their own labels. These are beautiful wines and a great new story to tell on the east side of Paso Robles.

CHATEAU MARGENE

4385 La Panza Road
Creston CA 93432
Phone: 805-238-2321
Fax: 805-238-2118
Web: www.chateaumargene.com
Email: info@chateaumargene.com
Hours: Saturday-Sunday noon-6 p.m. Friday;
Appt. with 72 hours advance notice.
Case production: 3,000

MADISON CELLARS

Jon and Margie Korecki are serious about their wine, but not too serious about themselves...and to that end, we will step out on a limb and quote from their web site blog on just how they got into the wine business.

"...The reason why we're in this business is because I like wine a whole lot and I have an ego the size of a house and think I can make better wine than anyone else with absolutely no experience. My wife is just along for the ride and the free wine..." It goes on in a similar direction for a while, an unselfconscious approach that is a breath of fresh air in the often-pretentious business of wine.

The Koreckis are originally from Connecticut. They came to Southern California about 13 years ago, veterans of the hi-tech wars; Jon dreaming of owning a vineyard and making wine. They purchased 50 acres in 2000, and have planted 30 of them so far. They planted most of the red Bordeaux and Rhone varietals. The vines are all own rooted. All their wines are estate bottled.

Jon is a minimal interventionist, very little fining or filtration. Primary fermentation for the reds is in open top fermenters, thrice-daily punch downs. He uses oak, but does not want it to overpower the wines... after fermentation, the reds see 20-40% new French and American oak, for 16-22 months. He prefers stainless steel or neutral oak for the whites, they make a beautiful Viognier. There is also a lovely rose of Syrah, just short of dry...Maximus is the name for their Bordeaux blend of Cabernet Sauvignon, Cabernet Franc and Merlot. It is their flagship wine, soon to be accompanied by Serendipity, a red Rhone blend.

The Koreckis are in the process of building an on site tasting room, forging ahead with national distribution of their wines, and "staying true to their roots" of growing and making classically styled Paso wines. We're looking forward to the fruits of their labors.

MADISON CELLARS
4540 Highway 41 West
Paso Robles, CA 93446
Phone 805-237-7544
Fax: 805-237-7798
Web: www.madisoncellars.com
Email: info@madisoncellars.com
Hours: By appointment only

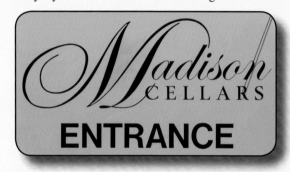

STILL WATERS VINEYARDS

In the mid-90's, Paul Hoover was cruising through northern San Luis Obispo county one day when he spotted Lee Alegre's (a well-known local vineyard consultant) team planting a new vineyard. Stopping by to speak with him, he started a conversation about installing a vineyard on one acre around the Hoover's house in Atascadero. A deal was struck, and one acre turned into close to sixty in a hurry, when they acquired a 58-acre parcel next door, already planted with Bordeaux varietals. There was a small house on the property, perfect for a tasting room, and a century old olive orchard. His home winemaking days were over.

They fine tuned the canopy and trellising systems to encourage much lower yields, and proceeded to graft over a few acres to Viognier, Pinot Grigio, Syrah, Malbec, Cabernet Franc, Sauvignon Blanc and Chardonnay The resulting fruit is in very high demand from the likes of Talley, Smith and Hook, Cinnabar, and Clos la Chance...they still sell 80 percent of their production. The winery on the property was bonded in 2003. The remaining fruit they estate bottle...he counts Gary Eberle and Steven Tebbe of Clos la Chance among his winemaking mentors. Paul uses stainless steel only for his tiny lots of Pinot Grigio and Sauvignon Blanc and the equally diminutive lots of Viognier and Chardonnay. They invented a giant "walk in freezer" room to ferment their whites in, which is unique. He uses a little neutral, or sometimes new, oak for limited (4-8 weeks) maturation only. Paul utilizes almost all French oak. They also produce a late harvest Sauvignon Blanc and a Rose of Syrah. There will be an as-yet unnamed white blend to match the red Reflection.

Their Cabernet Sauvignon is from the highest point on the ranch, grown in shaly soils that yield a paltry two tons per acre, gets a bit of a cold soak (as do all of his other reds) and 22 months of French and American oak. There is a Syrah, a Merlot, and Reflections, a Cabernet, Syrah, and Merlot blend. He regularly blends from his "spice rack" of six varietals that they grow, to round out flavors, textures and aromas. All his wines are handled minimally, utilizing gravity only to rack and move the wines. They are all small lots that sell out very quickly. Their mission statement is simple; "...be small, have fun and focus on quality..." born out by the fact that parts of the interview for this profile took place as Paul was about to disembark from a ski lift!

The produce a marvelous olive oil, too, from their 100 year old trees on the property. Paul's wife, Patty, does the books, and their daughter Stefanie, son Benjamin, brother-in-law Bill Elrod, and Bill's son Chris are all involved now with the winery and vineyards.

The winery sits atop a lovely hill, prettily landscaped, and complete with facilities for medium sized groups and is a marvelous place to visit, relax and enjoy the wonderful wines and view.

STILL WATERS VINEYARDS
2750 Old Grove Lane
Paso Robles, CA 93446
Phone: 805-237-9231
Fax: 805-438-3187
Web: www.stillwatersvineyards.com
Email: winery@stillwatersvineyards.com
Hours: Thursday-Monday 11 a.m.-5 p.m.
Case production: 1,500

CHAPTER TWELVE: SAN JUAN CREEK

> "WHAT THOUGH YOUTH GAVE LOVE AND ROSES,
> AGE STILL LEAVES US FRIENDS AND WINE."
> THOMAS MOORE

SAN JUAN CREEK (approximately 26,600 acres): warm region III-IV climate, more continental with only occasional marine influence this far inland, topography young river valleys with floodplain and alluvial terrace and fan deposits, north of La Panza Range, south of Cholame Hills, and immediately west of San Andreas Fault along San Juan Creek and its convergence into the Estrella River, younger alluvial soils with some older alluvial soils on high Estrella River terraces, semi-arid, sparse prairie (Elliott-Fisk, 2007).

Google Earth image (above) of the vineyards and surrounding lands in the Shandon area and San Juan Creek district of the Paso Robles AVA. Note that there are several different colors on the image indicating that it is a mosaic composite of airphotos taken at several different times during the years (brown or tan in the dry season and green in the wet season).

CHAPTER THIRTEEN: PASO ROBLES CANYON RANCH

Paso Robles Canyon Ranch (approximately 60,300 acres) warm region III-IV climate, most continental, but with pronounced cold air drainage in the evening reducing degree-day totals, older river terraces and alluvial fans immediately north of La Panza Range and west of San Andreas Fault, also bisected by Red Hills Fault, alluvial soils which are calcic and sodic soils in places in this semi-arid climate, vegetation semi-arid prairie/sage scrub with pines, oaks and sycamores in places in open woodlands (Elliott-Fisk, 2007).

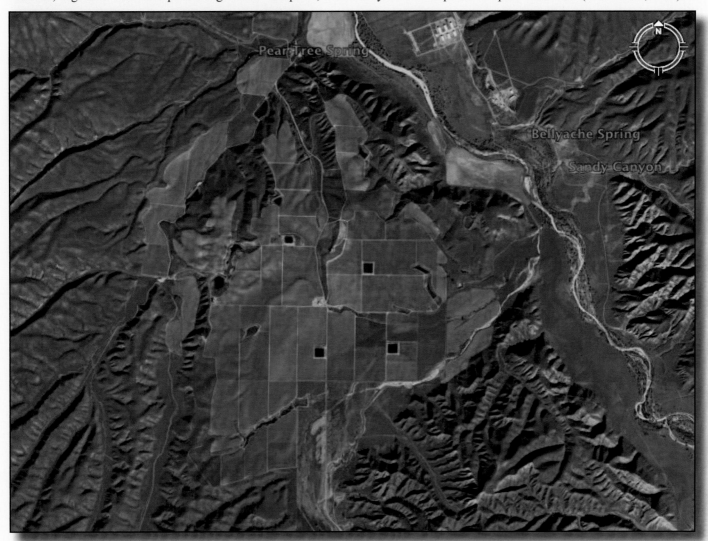

Google Earth image (above) of vineyards west of San Juan Creek and the surrounding lands in Paso Robles Canyon Ranch district.

CHAPTER FOURTEEN: WINERIES LOCATED OUTSIDE OF THE PASO AVA BOUNDARIES

CAYUCOS CELLARS

Stuart Selkirk is somewhat unique in this book. In a sea of transplants, he is a native son of the little seaside town of Cayucos. In the early 1980's, back when Paso Robles was just a gleam in a few peoples' eyes, he had the impudence to ask a neighbor what on earth was he doing with all those grapes in the back of his truck?

A couple of hours into his first crush, turning the handle of his neighbor's hand cranked grape crusher, he found out. Of course, then you have to punch down the caps, and then move the wine from the vats to the cellar. His Swiss neighbor had infected Stuart with a love of winemaking.

It was not long before a small vineyard was planted on his home property. Following his instincts about the winemaking process, he uses a minimalist approach; neutral oak, indigenous yeasts, and very little intercession. It's all Paso Robles fruit, along with some from his own property, some from the Templeton Gap, some from Adelaida Hills and a little from the east side of Paso Robles.

Their first production was in 1996, a tiny quantity...that has grown to 1,200 cases. I can't help but be impressed with the integrity of a winemaker whose only note listed for his 1999 Pinot Noir is a reference to the fifth amendment. Apparently, they were having difficulty coming up with a description. On a more lucid note, the 2000 Pinot Noir is all westside Paso fruit, the source for most of their wines. There is a "Devils Gate" Zinfandel, sourced from a vineyard that wished to remain anonymous, along with Zins from the Rio Seco and Rio Encino areas on the eastside. They produce a beautiful Syrah from eastside fruit, and a Chardonnay, fermented in neutral oak, from fruit grown in the Stacey Vineyard located in the Templeton Gap.

To complete their portfolio, they produce a wine called "Rustic One," a Cabernet/Zinfandel Blend made from the 1998 vintage, which spent 84 months in oak, and a luscious Zinfandel Port. Stuart and Laura Selkirk have added sons, Clay and Ross, with daughter Paige to the production lineup, for a truly family owned and operated property. This is just the sort of independent, freethinking, hard working spirit that exemplifies the Paso Robles spirit.

CAYUCOS CELLARS
143 N. Ocean Ave.
Cayucos, CA. 93430
Phone: 805-995 3036
Fax: 805-995-2415
Web: www.cayucoscellars.com
Email: wine@cayucoscellars.com
Hours: Winter, Wed.-Sun 11 a.m.-5 p.m..;
 Summer, Wed.-Mon. 11 a.m.-5 p.m.
Case production: 1,200

STEPHEN'S CELLAR (YORK MOUNTAIN AVA)

Steve Goldman and his father, Max, spearheaded the drive for the York Mountain AVA (an appellation east of the Paso Robles AVA boundary), and in 1983, when presented with the opportunity by the then BATF (now the TTB, for "Tax and Trade Bureau"), he jumped in and Stephen's Cellar was the first bonded winery in the newly minted region. York Mountain is a small, cool, tucked away mesoclimate on the westside of Paso Robles. He specializes in Pinot Noir, which thrives in this unusually moderate, foggy hillside pocket.

Goldman, and his wife, Lori planted his two acre vineyard in 2002, to, of course, Pinot Noir. The site is situated on a northerly slope, 1,500 feet above the Pacific. They practice organic and biodynamic viticulture. He espouses a minimum intervention philosophy in the winery, utilizing gravity and inert gas racking, with very little fining or filtration. His production is about 40 percent Paso Robles, 60 percent York Mountain, and as his own vines are still maturing, less than 10% is estate bottled.

Current releases of Pinot Noir include the vineyard designated Encell, Moore and the William Cain Vineyard. He uses some whole clusters, brief cold soaking, multiple daily hand punch downs, the free run juice sped off to 1-4 year old French oak barrels to mature with minimal racking for about 24 months. Press wine is kept separately. He also makes a McBride Vineyard Chardonnay, York Mountain AVA, made with much more attention to cooler fermentation in stainless steel, and a portion fermented and finished in one and two year old French and American barrel, with occasional lees stirring, for 9 months...a crisp, balanced approach.

Steve also makes a red blend called Rocky's, the current incarnation is a blend of Sangiovese, Barbera, Zinfandel, Petite Sirah, with a smattering of Alicante Bouchet (Central Coast appellation). He also makes a claret styled wine, of mostly Cabernets Sauvignon and Franc from the Carver vineyard on York Mountain, blended with Merlot form the Radike Vineyard in the Templeton Gap.

The Merlot is fermented in open top fermenters with indigenous yeast. There is a straight Carver Vineyard Cabernet Sauvignon as well, from this remarkable south facing vineyard at nearly 1,800 feet altitude. These are unique wines, from a unique mesoclimate in Paso Robles, well worth seeking out.

STEPHEN'S CELLAR AND VINEYARD
7575 York Mountain Road
Templeton, CA 93465
Phone/Fax: 805-238-2412
Web: www.stephenscellar.com
Email: steve@stephenscellar.com
Hours: by appointment only
Case production: 3,000

APPENDIX A

Case Study:
A vineyard soil survey report
for
Carmody McKnight Estate Wines

Carmody McKnight Vineyard Soil Survey Report

Introduction

Faculty and students of Polytechnic State University (Cal Poly), San Luis Obispo, CA conducted a project to inventory the soils and document the soil chemical and physical properties on chemically diverse soils at Carmody McKnight Vineyards.

The project began with a comprehensive soils study of the Carmody McKnight Vineyards in the Adelaida Hills district of San Luis Obispo County, CA, about seven miles northwest of Paso Robles. The primary objectives were to inventory and map the major vineyard soils and to provide laboratory characterizations of the important chemical and physical soil properties related to vineyard management.

There are two Cal Poly senior projects that resulted in a detailed soil survey and map of the Carmody McKnight Vineyards (Sloan and Westerling, 1996) and a comprehensive soil chemical property characterization of 19 soil profiles in the Carmody McKnight Vineyards (Roberts and Stubler, 1996).

Importance of Soils to Vineyard Management
Would you build a house without first having a plan and knowing the kinds of construction materials you would use? That is exactly what vineyard owners do when they plant rootstocks, choose grapevine scions, and implement a vineyard management system without thoroughly understanding the soils in which these plant materials will grow. Therefore, it is important to produce detailed soil maps prior to vineyard establishment and to analyze the important soil chemical and physical properties in order to make wise decisions about choice of grape rootstock and scion, fertilization management programs, cover crop choices, and irrigation system designs.

What do growers need to know about their vineyard soils in order to make good management decisions and where can they obtain the information that they need?

Sources of Soils Information
Soil survey reports produced by the United States Department of Agriculture (USDA) are widely available for much of the nation. These reports provide general soil maps and soil descriptions. However, for most intensive viticulture uses these reports provide only general reconnaissance information, which must be supplemented by detailed soil maps and complete site-specific chemical and physical soil characterizations. Earth and soil scientists working with private vineyard management companies produce these detailed soil maps.

Mapping of Soil Distribution and Variability
The first vineyard characterization task is to map the soil variability and soil distribution on the landscape. It has long been recognized that soils differ relative to variations in parent material, climate, landscape topography (relief), biosphere (plants and animals), and soil age. Soil variability on small vineyards is mostly determined by differences in soil parent materials and landform changes in slope steepness and aspect. The paragraphs that follow describe the soil mapping and sampling procedures that were used in this project and discusses those that are applied to other soil mapping efforts.

Until recently, soil mapping has been accomplished solely with the use of stereo aerial photographic base maps and topographic quadrangles. Relatively new remote sensing methods such as high-resolution aerial orthophotography, infrared aerial photography, and satellite imagery enable the soil mapper to detect relative soil water contents and other surface properties, which indicate soil variations. However, knowing soil properties to the depth of the root zone is important because it is the entire soil-root environment that ultimately determines the quality and management requirements for a given plant. Site base maps are produced with the use of these airborne technologies. However, none of the new remote sensing technologies are substitutes for excavating soil pits to determine soil variations with depth and sampling the soils for geochemical analyses.

Prior to the advent of modern geographic information systems (GIS) and the use of global positioning systems (GPS), the use of overlapping air photos viewed with stereoscopes were the tools allowing the soil scientist to see the land surface in three dimensions. Preliminary lines are drawn on the air photos to delineate areas of different topographic relief and aspect.

After preliminary air photo interpretation is complete, the soils within each polygonal delineation on the photo are then inspected on site along pre-determined transects and traverses to detect additional soil changes. Then final map lines, called soil map units, are made on the airphoto to represent the soil variations found on the landscape. Each soil map unit is named according to the dominant soil series, the surface soil texture, and the slope phase on which the soil is found. These map units vary in size from one to several tens of acres, depending on the size of the vineyard and the level of required detail.

Once the soil map unit boundaries are determined, a representative soil site for each soil map unit is selected for a complete morphologic description. A soil pit is excavated to at least two meters or to bedrock to expose a complete soil profile. The soil is described according to standard USDA nomenclature (Soil Survey Staff, 1994). The recognition and description of soil horizonation (layering) within soil pits is paramount to determining differences among soils. Soil horizons are layers in the soil that form as a result of sediment deposition or leaching of water through the soil profile over time. The described soils subsequently are classified according to soil taxonomy (Soil Survey Staff, 2006) and correlated to the series level using USDA official series descriptions. At the present time, over 20,000 different soil series are recognized in the United States.

After the soils description is complete, about two kilograms of soil are sampled from each major horizon (layer) and placed in plastic or paper bags. These soil samples are sent to a laboratory for complete characterization. In this study, soil samples from the major surface and subsurface horizons of each representative soil were analyzed in triplicate and average (statistical mean) chemical data are reported (Roberts and Stubler, 1996).

Carmody McKnight Vineyards
The project discussed in the remainder of this Appendix A began with a comprehensive soils study of the Carmody McKnight Vineyards in the Adelaida region of San Luis Obispo County, CA, northwest of Paso Robles. The primary objectives were to inventory and map the major vineyard soils using USDA historic soil maps and a new detailed soil map (Figure A1; Cal Poly soil map of Carmody McKnight vineyard) and to provide laboratory characterizations of the important soil chemical and physical properties.

Twenty one (21) soil pit were excavated to bedrock or two meters deep. Soil samples were collected and analyzed in the laboratory (Roberts and Stubler, 1996). After examination of the entire laboratory data set and the complete soil descriptions, it was determined that there are nine distinctly different soil types in the vineyard (Sloan and Westerling, 1996).

Vineyard Topography
The vineyard contains gently sloping rolling hills (Figure A1) generally running in a north south direction. Elevations range from 1,000 feet in the valleys to about 1,650 feet on the hillslope summits. Slopes range from 2 to 25 percent in the planted vineyard.

Vineyard Climate
The vineyard climate is a Mediterranean type which results in a cool, wet season in the winter and a dry, warm season in the summer. Rainfall ranges from 15 to 60 inches per year, occurring mostly between November and April. It is typically dry from May through October, so supplemental irrigation is used. Temperatures during the growing season range from about 40°F to over 100°F, with typical diurnal (day to night) fluctuations of 30°F to 50°F. The typical frost-free season runs from March through November. The soil moisture regime is classified as xeric and the soil temperature regime is classified as thermic (Soil Survey Staff, 2006; Table A1), as is the case throughout the entire Paso Robles AVA.

Vineyard Geology
The geology of both vineyards is reflected in two very differing lithologies (rock types) and resulting soil parent materials. The dominant geologic formation at the vineyards is mapped as the Monterey Formation of Tertiary-Miocene age. The dominant rocks at the vineyard are named as being the Sandholdt Member of the Monterey Formation.

The "Geologic Map of the Adelaida Quadrangle, San Luis Obispo, California" (Durham, 1968) describes the three primary Sandholdt Member rocks as chiefly mudstone, well indurated (hard, cemented), calcareous (contains calcium carbonate, a.k.a. lime), shaly (clay and silt sized particle grains), very pale orange or yellowish-gray, abundant middle Miocene (a geologic epoch about 19 to 26 million years ago) foraminifera (mostly microscopic plankton fossils and skeletons) and fish scales, fetid odor; Dolomitic carbonate rock (calcium- and magnesium-carbonate limestone), beds 1/2 to 2 feet thick, yellowish or grayish-brown in color.

In contrast to the sedimentary rocks, there are non-calcareous intrusive (coarse-grained) igneous rocks found mainly associated with the higher elevation hills and ridges at the vineyard that are largely uncultivated. The geologic map (Durham, 1968) describes these igneous rocks that are intrusive to the Sandholdt Member sedimentary rocks.

Most of the igneous rocks at the vineyard are mainly olivine "gabbro," which is a rock that is chemically similar to basalt but is more coarse-grained. That is, the minerals in the gabbro can be easily seen by the naked eye without the need for a microscope. The gabbro is reddish-gray in color and is weathered in most places to much redder-colored saprolite (weathered rock).

Vineyard Soils
The following discussion summarizes soil conditions at the Carmody McKnight vineyards determined initially by the Cal Poly team. More detailed soil descriptions can be found in the soil map unit description sections of a senior project report (Sloan and Westerling, 1996). Soils on the vineyards vary mostly according to topographic location and soil parent material (Table A3). Soils, such as Nacimiento, Diablo variant, and Linne variant, occur on upper hillslopes and narrow ridges tend to be shallower and contain more rock fragments. This is due to the fact that the harder, more weather-resistant rock types, such as olivine basalt, gabbro, limestone and brittle, partially metamorphosed (heated by magma) siltstone are located on the higher landscape positions. The Ayar, Capay, Cropley, and Zaca soils found on gently sloping, depositional landforms are deeper and more clayey (Table A3). These lower elevation soils

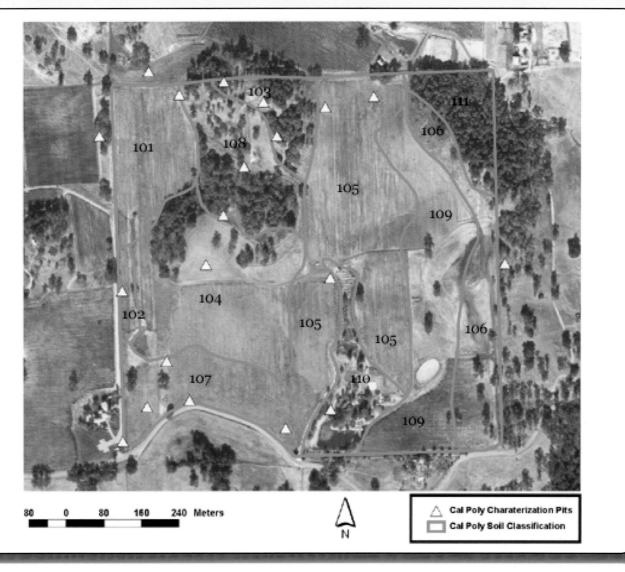

Figure A1: Soil map of Carmody McKnight Estate Vineyard

SOIL MAP LEGEND:

Soil map symbol: Soil map unit name	Soil Taxonomic Family
101: Ayar clay loam, 0 to 15 % slopes	Fine, smectitic, thermic Aridic Haploxererts
102: Ayar variant clay loam, 0 to 5 % slopes	Fine, smectitic, thermic Aridic Calcixererts
103: Linne variant sandy loam, 0 to 15% slopes	Fine-loamy, mixed, superactive, thermic Pachic Haploxerolls
104: Nacimiento loam, 0 to 15 % slopes	Fine-loamy, mixed, superactive, thermic Typic Calcixerolls
105: Cropley clay loam, 0 to 5 % slopes	Fine, smectitic, thermic Aridic Haploxererts
106: Diablo variant clay, 5 to 25 % slopes	Fine, smectitic, thermic Aridic Haploxererts
107: Capay clay loam, 0 to 5 % slopes	Fine, smectitic, thermic Aridic Haploxererts
108: Calodo variant clay loam, 5 to 15 % slopes	Fine-loamy, mixed, superactive, thermic Pachic Haploxerolls
109: Zaca clay, 5 to 10 % slopes	Fine, smectitic, thermic Vertic Haploxerolls
110: Developed Land	Soils not mapped.
111: Wildlife Habitat	Soils not mapped.

Table A1: Soil Map Units and Soil Classifications of the Carmody McKnight Vineyard Soils.

Soil map symbol: Soil map unit name	Taxonomic Soil Family
101: Ayar clay loam, 0 to 15 % slopes	Fine, smectitic, thermic Aridic Haploxererts
102: Ayar variant clay loam, 0 to 5 % slopes	Fine, smectitic, thermic Aridic Calcixererts
103: Linne variant sandy loam, 0 to 15% slopes	Fine-loamy, mixed, superactive, thermic Pachic Haploxerolls
104: Nacimiento loam, 0 to 15 % slopes	Fine-loamy, mixed, superactive, thermic Typic Calcixerolls
105: Cropley clay loam, 0 to 5 % slopes	Fine, smectitic, thermic Aridic Haploxererts
106: Diablo variant clay, 5 to 25 % slopes	Fine, smectitic, thermic Aridic Haploxererts
107: Capay clay loam, 0 to 5 % slopes	Fine, smectitic, thermic Aridic Haploxererts
108: Calodo variant clay loam, 5 to 15 % slopes	Fine-loamy, mixed, superactive, thermic Pachic Haploxerolls
109: Zaca clay, 5 to 10 % slopes	Fine, smectitic, thermic Vertic Haploxerolls
110: Developed Land	Soils not mapped.
111: Wildlife Habitat	Soils not mapped.

Table A2: Soil Types and Soil Forming Factors in the Carmody McKnight Vineyards.

Soil Type	Slope (%)	Slope Aspects	Soil Parent Materials
Ayar clay loam	0-15	south; west	alluvium from calcareous shale over residuum from calcareous mudstone and shale
Ayar variant clay loam	0-5	south	alluvium from calcareous shale and limestone over residuum from calcareous mudstone and shale
Calodo variant clay loam	5-15	south; west	alluvium and colluvium from calcareous shale over residuum from calcareous mudstone, shale and limestone
Capay clay loam	0-5	south	alluvium from calcareous shale over residuum from calcareous mudstone and shale
Cropley clay loam	0-5	south; west	alluvium from calcareous shale and gabbro over residuum from calcareous mudstone and shale
Diablo variant clay	5-25	south; west	alluvium from calcareous shale and mudstone over residuum from basalt and gabbro
Linne variant sandy loam	0-15	south; west	colluvium from basalt and gabbro over residuum from basalt and gabbro
Nacimiento loam	0-15	south	colluvium from limestone and calcareous shale over residuum from calcareous mudstone, shale and limestone
Zaca clay	5-10	south; west	alluvium from calcareous shale over residuum from calcareous mudstone and shale

are derived from a combination of finer textured (silts and clays) alluvium eroded from higher elevations and from calcareous shale that weathers to clay. Soils along major water drainageways, such as Ayar, Ayar variant, Capay, and Cropley, are formed from the combination of alluvium (water-transported sediment) and colluvium (gravity-transported sediment) overlying calcareous sedimentary rocks (Table A2). Soil properties vary across these map units resulting in soils located along the margins of the higher elevation igneous rock intrusions, which have properties that are mixtures of the weathered igneous and sedimentary rocks. The degree of influence that each different rock imparts to the soils depends on the soil's proximity to the source material. For example, the closer an alluvial soil is to an igneous rock intrusion, such as the Ayar soils, the greater the influence that igneous rock has on the resultant soil properties. Soils, such as Capay, are located along water drainages farther downslope from the igneous intrusions have slight chemical contributions from the weathered igneous rocks and mainly reflect the properties of the underlying sedimentary rocks.

Soils formed mainly from calcareous siltstone and shale (Ayar variant, Capay, Cropley, and Zaca soils) and soils formed from limestone and shale (Calodo and Nacimiento soils) are calcareous throughout have disseminated lime in the surface horizons and lime concentrations in soft masses, seams, and filaments in the subsurface horizons. These soils are alkaline and are relatively high in calcium, magnesium, copper, and sulfur (Tables A5 and A6). The amount of calcium and magnesium varies depending on the type of sedimentary rock. Calcitic limestone and calcareous mudstone and shale are higher in calcium, while dolomitic limestone is higher in magnesium. Copper is found in relatively moderate amounts in these sedimentary parent materials. Also, the sulfate-sulfur may be a remnant of the past use of ammonium sulfate fertilizers used in the historic production of barley on this land.

The Ayar soils formed from a mixture of igneous rock (olivine basalt, gabbro) alluvium and the underlying calcareous sedimentary rocks. When compared with the other soils derived from sedimentary rocks, they are lower in calcium and relatively high in iron and manganese. The Diablo variant and Linne variant soils formed from the igneous rocks (gabbro and basalt) are typically redder or more orange in color. They are slightly acid to neutral, relatively high in iron, manganese, potassium, and zinc; and relatively low in calcium, magnesium, and sulfur (Tables A5 and A6). Copper and sodium are also present in small amounts. Soil organic matter levels range from two to four percent in the topsoils and one to two percent in the subsoils (Tables A5 and A6). These relatively high organic matter levels are a reflection of the past and present oak woodland-grassland savannah ecosystem and agricultural practices. Organic matter decomposition processes result in low but relatively consistent levels of nitrogen and phosphorus being added to the soil solution.

The soils are classified as Mollisols and Vertisols (Soil Survey Staff, 2006; Table A1). Mollisols are soils of native grasslands that have thick dark fertile topsoils. Vertisols are soils with greater than 30 percent expansive clays (called smectite or montmorillonite) throughout their upper profiles. The dominant surface and subsurface soil textures are clays and clay loams (Table A3). Most of the soils have diagnostic subsurface horizons with illuvial clay and/or calcium carbonate (i.e., argillic and calcic horizons) indicating past soil development (Tables A3 and A4). Some of these soils have illuvial horizons, (subsoil layers having an accumulation of layer-lattice silicate clay minerals, mostly montmorillonite, and carbonate accumulations), resulting in soils with relatively high subsoil clay contents. These montmorillonite clay soils have relatively high cation exchange capacities.

Summary

In order to effectively select a vineyard site and wisely manage a vineyard, detailed soil survey reports should be produced to supplement any existing, less detailed USDA soil maps. Detailed soil survey reports and maps, along with characterization of the major soils' chemical and physical properties are used by vineyard managers to help make wise decisions about grape rootstocks and scions, cover crop choices, vineyard fertilization programs, and irrigation system designs.

Future studies, begun with the 2003 harvest season, focused on soil effects on the quality of Cabernet Sauvignon wine grapes based on the site *terroir* (variations in soil types and microclimates) within the Carmody McKnight Vineyards. These studies are in progress and the complete research results will be published in 2008.

Table A3: Summary of Representative Soil Properties in the Carmody McKnight Vineyards (see the detailed soil descriptions in Sloan and Westerling, 1996).

Soil Type	Soil Horizon(s)	Soil Horizon Depths (cm)	Texture or weathered rock type	Rock fragment (%)	Soil Structure Type
Ayar clay loam	Ap	0-15	clay loam	10	blocky
	Btss	15-130	clay	10	prismatic
	Cr	130-150	calcareous shale	95	rock
Ayar variant clay loam	Ap, A	0-30	clay loam	5	blocky
	Btss	30-80	clay	5	prismatic
	Bk	80-150	clay loam	10	massive
Calodo variant clay loam	A1, A2	0-75	clay loam	10	granular, blocky
	A3	75-95	gravelly clay loam	15	blocky
	Cr	95-150	calcareous shale, limestone	95	rock
Capay clay loam	Ap, A1, A2	0-50	clay loam	10	granular, blocky
	Btss	50-140	clay	10	prismatic
	Cr	140-150	calcareous shale	95	rock
Cropley clay loam	Ap, A	0-35	clay loam	5	blocky
	Btkss	35-85	clay	10	prismatic
	Cr	85-150	calcareous mudstone, shale	95	rock
Diablo variant clay	A1, A2	0-50	clay	5	blocky
	Bw	50-80	clay	10	prismatic
	Cr	80-150	basalt, gabbro	95	rock
Linne variant sandy loam	A1	0-17	very gravelly sandy loam	40	granular
	Bw	17-75	very gravelly sandy loam	40	blocky
	Cr	75-150	basalt, gabbro	95	rock
Nacimiento loam	A	0-35	loam	10	granular
	Bk	35-80	gravelly clay loam	25	blocky
	Cr	80-150	calcareous shale; limestone	95	rock
Zaca clay	Ap	0-10	clay	10	granular
	Bw	10-90	clay	10	blocky
	Cr	90-150	calcareous mudstone; shale	95	rock

Table A4: Summary of Representative Soil Properties in the Carmody McKnight Vineyards (see the detailed soil descriptions in Sloan and Westerling, 1996).

Soil Type	Soil Horizon(s)	Soil Horizon Depths (cm)	Dominant Moist Color (Munsell)	Dominant Moist Color (Descriptive)	Calcareous (yes/no)
Ayar clay loam	Ap	0-15	10YR 3/3	dark brown	yes
	Btss	15-130	10YR 3/3	dark brown	yes
	Cr	130-150	10YR 5/4	yellowish brown	yes
Ayar variant clay loam	Ap, A	0-30	10YR 3/1	very dark gray	yes
	Btss	30-80	10YR 3/1	very dark gray	yes
	Bk	80-150	10YR 5/4	yellowish brown	yes
Calodo variant clay loam	A1, A2	0-95	10YR 3/2	very dark grayish brown	yes
	A3	75-95	10YR3/3	dark brown	yes
	Cr	95-150	10YR 5/8	yellowish brown	yes
Capay clay loam	Ap, A1, A2	0-50	10YR 3/2	very dark grayish brown	yes
	Btss	50-140	10YR 3/2	very dark grayish brown	yes
	Cr	140-150	10YR 5/4	yellowish brown	yes
Cropley clay loam	Ap, A	0-35	10YR 2/1	black	no
	Btkss	35-85	10YR 3/1	very dark gray	yes
	Cr	85-150	10YR 5/8	yellowish brown	yes
Diablo variant clay	A1, A2	0-50	10YR 3/3	dark brown	no
	Bw	50-80	10YR 4/4	dark yellowish brown	no
	Cr	80-150	7.5YR 5/8	strong brown	no
Linne variant sandy loam	A1	0-17	10YR 3/3	dark brown	yes
	Bw	17-75	10YR 3/3	dark brown	yes
	Cr	75-150	7.5YR 5/8	strong brown	yes
Nacimiento loam	A	0-35	10YR 4/3	brown	yes
	Bk	35-80	10YR 5/3	brown	yes
	Cr	80-150	10YR 6/6	brownish yellow	yes
Zaca clay	Ap	0-10	10YR 3/1	very dark gray	yes
	Bw	10-90	10YR 3/1	very dark gray	yes
	Cr	90-150	10YR 5/4	yellowish brown	yes

Table A5: Surface Soil Chemical Results from Representative Soil Profiles within Carmody McKnight Vineyard. (most data are expressed in ppm, on a dry soil basis, except OM, as % by weight).

Map Unit	Soil Type	Ca	Mg	Na	K	Fe	Cu	Mn	Zn	P[1]	SO$_4$	N[2]	OM[3]
						ppm							%
101	Ayar clay loam	3965	1112	214	472	46	2.84	23	2.01	4.45	2.89	74	3.69
102	Ayar variant clay loam	5104	1817	213	429	20	2.36	36	2.06	2.60	7.80	78	3.90
103	Calodo variant clay loam	3796	967	219	483	20	1.22	17	1.75	2.66	1.40	75	3.73
104	Capay clay loam	5034	331	198	431	5	6.62	8	1.69	4.53	2.40	70	3.49
105	Cropley clay loam	5796	1043	276	496	25	3.51	7	1.20	2.43	13.29	60	2.98
106	Diablo variant clay	4283	1600	379	400	59	1.88	12	0.84	1.60	2.60	60	3.00
107	Linne variant sandy loam	5368	642	210	462	19	4.95	14	1.81	4.08	2.25	54	3.05
108	Nacimiento loam	5002	717	225	696	11	4.56	10	1.95	3.09	1.59	76	3.80
109	Zaca clay	4875	996	171	750	8	3.34	9	0.88	2.30	0.98	52	2.60
	Ave.	4803	1025	234	513	24	3.48	15	1.58	3.08	3.91	66	3.36
	Std. Dev.	659	460	61	123	18	1.68	9	0.47	1.04	4.05	10	0.46

FOOTNOTES: [1] Available phosphorus. [2] Total nitrogen. [3] Organic matter.

Table A6: Subsurface Soil Chemical Results from Representative Soil Profiles within Carmody McKnight Vineyard. (most data are expressed in ppm, on a dry soil basis, except OM, as % by weight).

Map Unit	Soil Type	Ca	Mg	Na	K	Fe	Cu	Mn	Zn	P[1]	SO$_4$	N[2]	OM[3]
						ppm							%
101	Ayar clay loam	3897	1372	229	296	60	2.87	13	1.15	2.13	2.20	32	1.59
102	Ayar variant clay loam	5575	1933	225	179	8	1.64	7	0.64	1.18	0	50	2.51
103	Calodo variant clay loam	2354	833	204	758	19	0.92	75	2.42	0	0	32	1.60
104	Capay clay loam	4990	350	213	288	9	5.85	4	1.10	4.79	0.85	28	1.39
105	Cropley clay loam	5713	1316	274	383	23	3.14	3.2	0.64	1.96	11.85	34	1.68
106	Diablo variant clay	4808	1663	538	204	55	1.80	7	0.70	7.25	1.54	21	1.06
107	Linne variant sandy loam	4823	1135	313	531	17	3.95	12	0.78	2.25	0.75	46	2.30
108	Nacimiento loam	4900	1154	246	321	5	3.30	4	0.34	1.18	1.20	12	1.22
109	Zaca clay	4900	1154	246	321	5	3.30	4	0.34	1.18	1.20	24	1.22
	Ave.	4662	1212	276	365	22	2.97	14	0.90	2.44	2.18	31	1.62
	Std. Dev.	1007	455	104	179	21	1.45	23	0.64	2.23	3.69	12	0.49

FOOTNOTES: [1] Available phosphorus. [2] Total nitrogen. [3] Organic matter.

APPENDIX B

Important Soil Considerations and
Information for
Wise Vineyard Management

• Soil Texture
• Soil and Rootstock Relationships
• Plant Essential Nutrients for Grapes
• Bibliography

Importance of Soil Texture to Vineyard Management (Rice, 2002)

What is Soil Texture?

Soil texture refers to the weight proportion of the mineral soil separates for particles less than 2 millimeters (mm) as determined from a laboratory particle-size distribution. Soil texture is the most important physical property which influences water holding capacity, root growth and overall vine vigor. Natural soils are comprised of soil particles of varying sizes. The soil particle-size groups, called soil separates, are sands (the coarsest), silts, and clays (the smallest). The terminology used to describe soil texture in the popular press often does not adhere to the standard definitions. Therefore, much confusion and outright misinformation is spread about this vineyard soil property. The purposes of this section are to present the accepted standardized definitions for soil texture terms and to discuss the importance of soil texture to vineyard management.

Soil Separate Sizes

The U.S. Department of Agriculture (USDA) has established limits of variation for the soil separates and has assigned a name to each size class (Table B1). This system has been approved by the Soil Science Society of America (SSSA) and is the one used in all published USDA soil survey reports. Most of the pertinent USDA publications regarding soil survey standards and soil texture terms can be accessed from USDA internet page cited in the References section of this article. Other standard soil science terminology can also be referenced by examining the Glossary of Soil Science Terms linked to the SSSA internet page (SSSA, 2001).

USDA Soil Textural Classes

The texture classes (shown on the USDA textural triangle; Figure B1) are sand, loamy sands, sandy loams, loam, silt loam, silt, sandy clay loam, clay loam, silty clay loam, sandy clay, silty clay, and clay.

Subclasses of sand are subdivided into coarse sand, sand, fine sand, and very fine sand. Subclasses of loamy sands and sandy loams that are based on sand size are named similarly. Detailed definitions for the many different subclasses of sands, loamy sands, and sandy loams are found in the USDA soil survey manual (Soil Survey Division Staff, 1993).

The textural triangle is used to resolve problems related to the detailed word definitions (Figure B1). The eight subclasses in the sand and loamy sand groups provide refinement greater than can be consistently determined by field techniques. Only those textural distinctions that are significant to agricultural use and management and that can be consistently made in the field are commonly applied.

Groupings of soil texture classes

It is often convenient to speak of general texture groups. The general soil texture groups, in three (sandy, loamy or clayey) or five classes, are outlined in Table B2. In some areas where soils are high in silt, a general class, silty soils, is used for silt and silt loam.

Determination of Soil Texture

Apparent field texture is a tactile evaluation only with no inference as to laboratory test results. Field estimates of soil texture should always be checked against laboratory determinations and the field criteria should be adjusted as necessary. Sand particles feel gritty and can be seen individually with the naked eye. Silt particles cannot be seen individually without magnification; they have a smooth feel to the fingers when dry or wet.

Clay soils range from slightly sticky to very sticky, when wet. This is because soil texture and soil clay mineralogy are not directly related. Soils dominated by smectitic (swelling and cracking) clays are more sticky and more plastic than soils that contain similar amounts of micaceous (high in muscovite or biotite mica) or kaolinitic (non-swelling) clays.

Soil texture and soil organic matter content are also not directly related. In fact, to determine soil texture using standard USDA methods, the organic matter and all other soil aggregating agents (like calcium carbonate, silica and iron oxides) are chemically removed from the total soil sample prior to the determination of soil texture. Therefore, standard soil texture methods call for removal of these most chemically important soil components. This fact makes it very important to look at the soil chemical properties like pH, organic matter content, and plant available nutrients, in addition to the soil texture, to make wise viticultural management decisions.

Many analytical labs will determine the soil saturation percentage (i.e., the amount of water, by weight, in a saturated soil sample) and make indirect soil texture determinations from this value (Table B3). Generally, the higher the saturation percentage, the higher total soil clay and organic matter contents. The saturation percentage is directly related to the total soil porosity and total soil water-holding capacity and, therefore, is a valuable number to use to aid irrigation system design.

Rock Fragments

Rock fragments are unattached pieces of rock, which are 2 mm in diameter or larger that are hard or strongly cemented. They are physically removed (by sieving) from the soil separates in the laboratory determination of soil texture. Then, volume and weight measurements are performed to determine their amounts in soil.

Rock fragments are described by size, shape, and, in some cases, the type of rock. The rock fragment classes are pebbles, cobbles, channers, flagstones, stones, and boulders (Table B4). If one size or range of sizes predominates in a soil, the textural class is modified using additional information. For example, compound terms like "fine pebbles," "cobbles, 100 to 150 mm in diameter," "channers, 25 to 50 mm in length" may be used.

Gravel is a collection of pebbles that have diameters ranging from 2 to 75 mm. The term is applied to the collection of pebbles in a soil layer with no implication of geologic rock type. The terms "pebble" and "cobble" are usually restricted to rounded or subrounded fragments; however, they can be used to describe angular fragments if they are not flat.

Words like granite, limestone, and shale refer to a rock type, not a rock fragment. The composition of the fragments can described as follows: "granite pebbles," "limestone channers" and "shale gravels." The upper size of gravel is 75 mm (3-inch). This coincides with the upper limit used by many engineers for grain-size distribution computations. The 5 mm and 20 mm divisions for the separation of fine, medium, and coarse gravel approximately coincide with the sizes of openings in the "number-4" screen (4.76 mm) and the "3/4-inch" screen (19.05 mm) used by engineers.

The 75 mm (3-inch) limit separates gravel from cobbles. The 250 mm (10-inch) limit separates cobbles from stones, and the 600 mm (24 inch) limit separates stones from boulders. The 150 mm (channers) and 380 mm (flagstones) limits for thin, flat fragments follow conventions used for many years to provide class limits for plate-shaped and crudely spherical rock fragments that have about the same soil use implications as the 250 mm limit for spherical shapes.

In USDA soil survey reports, the adjective describing rock fragments in soils is used as the first part of the textural class name according to the following conventions:

- less than 15 percent by volume. No mention of rock fragments is used.
- 15 to 35 percent by volume. The dominant kind of rock fragment is used (e.g., "gravelly loam," "cobbly clay loam").
- 35 to 60 percent by volume. The word "very" precedes the name of the dominant kind of rock fragments (e.g., "very cobbly loam" "very cobbly sandy loam").
- > 60 percent by volume. Add the word "extremely" in front of the coarse fragment name (e.g., "extremely gravelly loam" "extremely cobbly clay loam").
- If there are too few soil separates present to determine the soil textural class (less than 10 percent by volume), terms such as "gravel" "cobbles," "stones," or "boulders" are used, as appropriate, without mention of the textural class.

Soils generally contain rock fragments smaller or larger than those identified by the adjective term. The rock fragment adjective applies to the most predominant rock fragment size found within any soil layer. For example, on a volume basis, a stony loam with 20 percent stones may also contain 10 percent gravel-sized pebbles, but "gravelly" is not mentioned in the name. In this case, stones are the most abundant rock fragment size and "stony" is used as the adjective.

More precise estimates of the amounts of rock fragments than are provided by the defined classes are needed for some purposes. If more precise information is needed, estimates of percentages of each size class or combination of size classes are included in the description: such as "very cobbly loam; 30 percent cobbles and 15 percent gravel" or "silt loam; about 10 percent gravel."

If rock fragments are significant in soil vineyard management, they are the basis for designing and describing new soil map units. These soil map units are identified on the vineyard soil maps. In contrast, bedrock exposed at the earth's surface is not soil and is separately identified on soil maps.

The volume occupied by individual rock fragments can be seen in the field and aggregate volume percentage can be estimated. Rock fragment volume percentage may be converted to a weight percentage, using USDA conversion tables or laboratory measurements.

Influence of Soil Texture on Other Soil Properties

As discussed earlier, individual soil mineral particle diameters range over six orders of magnitude, from boulders ($\geq$600 mm) to clays (less than 10^{-9} mm) which can only be seen with an electron microscope. The soil textural classification established by the USDA is used for agricultural soils. The size ranges for these separates are not purely arbitrary, but reflect major changes in how the particles behave and in the physical properties they impart to soils (Table B5).

Gravels, cobbles, boulders and other rock fragments which are >2 mm in diameter will affect soil behavior, but they are not considered to be part of the fine earth fraction (soil particles <2 mm) to which the soil texture term applies. As described earlier, an adjective term is used with the soil textural class to identify the rock fragment amounts in soil.

Texture is an important soil characteristic because it will, in part, determine water intake rates (infiltration), water movement through soil (hydraulic conductivity), soil water holding capacity, the ease of tilling the soil, the amount of aeration (vital to root growth), and will influence soil fertility.

For instance, a coarse sandy soil is easy to till, has plenty of aeration to stimulate root growth, and is easily irrigated. However, this same sandy soil will rapidly dry out after irrigation due to its low water holding capacity. Water soluble plant nutrients (like nitrates and potassium) will be rapidly leached below the vine root zone by percolating waters.

In contrast, moist clay soils (over 35 percent clay) are composed of very small particles that fit tightly together with fewer large interconnected pores. Clay soils should be irrigated less frequently than sands, but with higher amounts of water and over longer periods of time. Clay soils have greater cation exchange capacities (CEC) and will adsorb higher amounts of water-soluble plant nutrients (especially potassium, calcium and magnesium). Wet clay soils are also difficult to till due to their relative stickiness and inability to support the weight of a tractor.

Summary

For most intensive vineyard management operations, a detailed soil map should be prepared to supplement the more general USDA soil survey reports. The soil textures for each soil layer within one meter should be identified for the dominant soils using the standard USDA terms defined in this paper. When describing soil textures within a vineyard, it is important to not only identify the soil textural classes but also the soil parent material rock types and the amounts of rock fragments within the soil profiles. Related soil properties like saturation percentage should also be determined to aid in irrigation system design.

Soil texture information should also be supplemented with data regarding soil organic matter content and soil chemical properties, such as pH and plant essential nutrient concentrations. Soil organic matter and humus contents (highly decomposed organic matter) will modify the general effects of soil mineral texture (Table B5) by causing soil structure formation, increasing soil water holding capacity and increasing CEC. The interrelationships among all these soil properties should be considered when vineyard management decisions are made.

Table B1. Soil Separates and Their Diameter Ranges

Soil Separate	Diameter (mm)	Visual Size Comparison of Maximum Size
Sand	2.00 to 0.05	
Very coarse	2.00 to 1.00	House key thickness
Coarse	1.00 to 0.50	Small pinhead
Medium	0.50 to 0.25	Sugar or salt crystals
Fine	0.25 to 0.10	Thickness of book page
Very fine	0.10 to 0.05	Invisible to naked eye
Silt	0.05 to 0.002	Visible under light
Coarse 0.05 to 0.002	microscope	
Medium	0.02 to 0.005	
Fine	0.005 to 0.002	
Clay	<0.002	Visible with an electron
Coarse 0.002 to 0.0002	microscope	
Fine	<0.0002	

Table B2. General Soil Texture Groups

General Texture Group Terms	Texture Classes
Sandy soil materials:	
Coarse textured	Sands (coarse sand, sand, fine sand, very fine sand). Loamy sands (loamy coarse sand, loamy sand, loamy fine sand, and loamy very fine sand.
Loamy soil materials:	
Moderately coarse-textured	Coarse sandy loam, sandy loam, fine sandy loam.
Medium-textured	Very fine sandy loam, loam, silt loam, silt.
Moderately fine-textured	Clay loam, sandy clay loam, silty clay loam.
Clayey soils:	
Fine-textured	Sandy clay, silty clay, clay

Table B3. Saturation Percentage and Approximate Soil Texture Classes

Soil Saturation Percentage	Approximate Soil Texture Classes
Below 20	Sand or Loamy Sand
20 - 35	Sandy Loam
35 - 50	Loam or Silt Loam
50 - 65	Clay Loam
65 - 150	Clay
Above 150	Usually Organic soils (peat or muck)

Table B4. Terms for Describing Rock Fragments in Soils

Shape* and size	Noun	Adjective
Spherical, cubelike, or equiaxial:		
2-75 mm diameter	**Gravel (pebbles)**	**Gravelly**
2-5 mm diameter	Fine	Fine gravelly
5-20 mm diameter	Medium	Medium gravelly
20-75 mm diameter	Coarse	Coarse gravelly
75-250 mm diameter	**Cobbles**	**Cobbly**
250-600 mm diameter	**Stones**	**Stony**
≥600 mm diameter	**Boulders**	**Bouldery**
Flat:		
2-150 mm long	Channers	Channery
150-380 mm long	Flagstones	Flaggy
380-600 mm long	Stones	Stony
>600 mm long	Boulders	Bouldery

(Footnote)

*The roundness of the fragments may be indicated as angular (strongly developed faces with sharp edges), irregular (prominent flat faces with incipient rounding of corners), subrounded (detectable flat faces with well-rounded corners), and rounded (flat faces absent or nearly absent with all corners).

Table B5. Generalized Influence of Soil Separates on other Soil Properties and Behavior

Property/Behavior	Sand	Silt	Clay
Water-holding capacity	Low	Medium to High	High
Aeration, when moist	Good	Medium	Medium to Poor
Hydraulic (water) conductivity	High	Slow to Medium	Slow to very slow
Soil organic matter level	Low	Medium to High	High to Medium
Decomposition of organic matter	Rapid	Medium	Slow
Warm-up in spring	Rapid	Moderate	Slow
Compactability	Low	Medium	High
Susceptibility to wind erosion	Moderate	High	Low (unless fine sand)
Susceptibility to water erosion	Low	High	Low if aggregated, High if not
Shrink-swell potential	Very low	Low	Moderate to very high
Sealing of ponds and dams	Poor	Poor	Good
Suitability for tillage when wet	Good	Medium	Poor
Pollutant leaching potential	High	Medium	Low (unless cracked)
Cation exchange capacity (CEC)	Low	Medium	High
Resistance to pH change	Low	Medium	High

Figure B1. USDA Textural Triangle

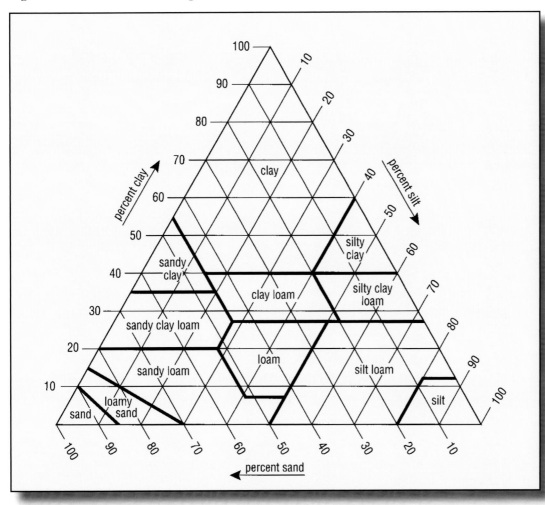

Soil Considerations for Wise Wine Grape Rootstock Choices

Matching the best rootstock to the soil type is a challenge involving many variables. Knowledge of rootstock disease resistance and soil pH (acid vs. alkaline) should be balanced with a knowledge of soil and root vigor potential in order to make the appropriate rootstock choices (Christianson, 2003; Walker, 2006).

When choosing a rootstock, the first factor to be considered is its resistance to phylloxera and nematodes. Using a rootstock that is phylloxera resistant is imperative in California, since nearly all soils contain phylloxera and past infestations have had devastating effects in California and Europe (Coombe and Dry, 1992a). Nematode resistance depends on the species of nematode. The two primary nematodes that pose a threat in California vineyards are root knot and Dagger. Root knot can be defended against using a variety of rootstocks, but there are only a few rootstocks resistant to Dagger. It is important to test for soil nematodes before choosing a rootstock (Walker, 2006).

The other factors that should be considered when choosing a rootstock include soil depth, available water holding capacity (AWHC), soil texture, soil vigor (related to water supply and nutrient levels), future water availability, lime (calcium carbonate) content, and soil salinity (Wolpert et al., 1992). Soil depth, AWHC, plant nutrient levels (especially nitrogen) and soil texture are all considered to determine soil vigor. High vigor soils should be planted with a low vigor rootstocks to ensure the optimal grape quality and proper fruit ripening (Derlas and Pouget, 1979).

When looking at future water availability, irrigation water supply and land slope and aspect should be taken into consideration. South facing slopes (in the northern Hemisphere) are generally warmer and, therefore, often require more frequent irrigation water additions. Drought resistant rootstocks should be planted on these south slopes and in shallow soil areas (Carbonneau, 1985). Drought resistance should always be taken into consideration in a semi-arid climate, such as the Paso Robles AVA.

Calcareous soils with high lime levels (alkaline soil pH) should be planted with lime-tolerant rootstocks (Pouget and Delas, 1989). If these lime tolerance is not considered when choosing a rootstock, problems with the grape vines and fruit quality may develop due to metal micronutrient deficiencies.

Tables C1 and C2 are examples that list grape rootstock properties and rootstock choices at Stephan Vineyards and L'Aventure Winery, where the soils are derived from both calcareous (alkaline soil pH) and siliceous (acid soil pH) parent materials (Rice et al., 2007). Complete discussions of grape rootstocks and soil interactions are available in the following references (Christianson, 2003; Hardie and Cirami, 1992; Walker, 2006; Wolpert et al., 1992).

Rootstock Evaluation at Stephan Vineyard (Rice et al., 2007)

As an example of rootstock selection and evaluation, the rootstocks at the Stephan Vineyard were examined relative to soil type (Rice et al., 2007; Tables C1 and C2). Overall, the existing rootstocks at Stephan Vineyard appropriately match the soil characteristics. The 110R, 1103P or 3309C rootstocks should rarely be used on highly calcareous (high lime) soils since they all have moderate lime tolerance. They are best used on soils with low to moderate lime levels. If the soil is not acidified, Fe, Zn and Mn deficiencies may occur (Walker, 2006). Rootstocks 3309C and 420A result in low vigor scions. When planted on shallow, low vigor soils such as Calleguas or Calodo, the grape vines may not grow vegetatively to their optimum. The Fercal and 140Ru rootstocks are good choices for planting in the highly calcareous soils, since they has medium vigor potential with a high degree of lime tolerance (Christianson, 2003; Pouget and Delas, 1989).

Table C1: Existing rootstocks and their properties at Stephan Vineyard (Christianson, 2003; Rice et al., 2007)

Rootstock	Phylloxera Resistance	Root Knot Resistance	Drought Tolerance	Salinity Tolerance	Lime Tolerance	Vigor Potential Influence on Scion	Mineral Nutrition Influence on Scion	Soil Adaptation & other characteristics
110R (Richter)	High	Low-Medium	High	Medium	Medium	Medium	N: med. P: high K: low-med. Mg, Zn: med.	Hillside soils; acid soils. Develops slowly in wet soils
140Ru (Ruggeri)	High	Low-Medium	High	Medium-High	Medium-High	High	N: med.-high P, Mg: high K: low	Adapted to drought and acid soil. Does poorly in non-irrigated low K soils
1103P (Paulsen)	High	Medium-High	Low-Medium	Low-Medium	Low-Medium	Medium-High	N: med.-high P, Mg: high K, Zn: low-med.	Adapted to drought and saline soils
161-49C (Couderc)	High	Low	Low	Low	High	Low	N: low K: med.-high	Good for high density plantings
3309C (Couderc)	High	Low	Low-Medium	Low-Medium	Low-Medium	Low-Medium	N: med.-high P, Ca: low K, Mg, Zn: med.	Deep soils; sensitive to latent viruses and tolerant of cold injury
420A Millardet et de Grasset)	High	Medium	Low-Medium	Low	Medium-High	Low	N, P, K: low Mg: med. Zn: low-med.	Fine textured; fertile soils. Scions tend to overbear when young.

Table C2: Soil properties and existing rootstocks at Stephan Vineyard (Rice et al., 2007)

Soil Series	Soil Textures	Avail. Water Hold. Capacity	Salinity Level	Lime Level	Soil Vigor Potential	Existing Rootstocks
Botella	clay loam	High	Low	High	High	110R, 140Ru
Calleguas	clay loam	Very low	Low	High	Low	110R, 140Ru, 3309C
Calodo	loam	Low	Low	High	Low	110R, 140Ru, 3309C, 420A
Gazos	sandy clay loam; clay loam	Low	Low	None	Medium	110R, 1103P, 140Ru, 161-49C, 3309C, 420A
Linne	clay loam	Low	Low	High	Medium	110R, 1103P, 3309C, 420A
Lockwood	clay loam; clay	Moderate	Low	None	High	110R, 140Ru, 161-49C, 3309C
Lopez	gravelly silty clay loam	Very low	Low	None	Low	1103P
Norrena	clay loam	Low	Low	Low	Low	140Ru
Windage	silt loam; silty clay loam	High	Low	None	High	110R, 3309C, 420A

Roles of the Plant Nutrients and their Requirements for Grapes

Soil tests should be conducted to characterize the soil chemical environment prior to vineyard establishment. After vineyard establishment, soil tests may periodically be conducted when soil-related nutrient problems or soil chemical toxicities (like soluble salts or boron) are suspected. Typically, visual examinations of grape leaf blades and overall vine growth, along with routine leaf blade and/or petiole testing, are conducted in established vineyards in order to detect nutrient deficiencies. A combination of nutrient management strategies, including foliar nutrient sprays, liquid fertigation applications in the irrigation water, and direct soil applications are used to correct nutrient deficiencies in vineyards. Excellent internet access to viticulture information related to fertilizer recommendations and irrigation management is now available online (Univ. of California, 2007).

The following Table D1 and discussion of the important plant essential nutrients will serve as a practical reference to understand their roles in viticulture. A professional agriculturist should be consulted to provide solutions when chronic nutrient-related problems are detected in a vineyard.

Discussion of the Plant Essential Nutrients Important in Viticulture

Nitrogen (N)

Nitrogen is a primary component of plant proteins, nucleic acids, and chlorophyll. Nitrogen (N) is the most commonly deficient element in crop production (Havlin et al., 2005), but not in mature wine grape vineyards. An excessive amount of N adversely affects fruit production by stimulating heavy vegetative growth and reduced fruit production (Peyrout des Gachonset al., 2005; Robinson, 1992). Organic matter and a variety of fertilizers supply nitrogen to the soil solution. Deficiency symptoms include reduced vigor of leaves and shoots, and overall yellowing or chlorosis of the entire leaf and other green tissues. Grapes require an estimated 75 pounds N per acre for stem, leaf, and fruit growth and bud fertility (Christensen, 1990; Coombe and Dry, 1992). Following bloom, the nutrient demand in the fruit greatly increases and the emphasis in fertilizer nutrients shifts slightly from nitrogen to potassium (Grant, 2002)

Phosphorus (P)

Phosphorus is essential to early plant growth and root formation (CFA, 1998). It plays an important role in energy storage and transfer. Phosphorus is also a component of nucleoproteins (DNA and RNA) and phospholipids (Mullins et al., 1992). Most plants require 0.0003 to 0.3 ppm of phosphorus (P) for a healthy life cycle (Havlin et al., 2005). Most soils contain only 0.02 to 0.10% P, which is supplied by weathered minerals, fertilizers and organic matter (Tisdale et al. 1985). A P-deficiency is exhibited as a gradual reduction in shoot growth sometimes without severe leaf symptoms, basal leaf yellowing or turning pale and red dots may form near edges of mid or terminal lobes of basal leaves which later coalesce forming red bars at right angles to the vein (Coombe, 1992).

Potassium (K)

Potassium encourages root growth, fruit production, and a plant's resistance to disease (CFA, 1998). It functions in the sugar translocation, starch formation, and the opening and closing of the stomata. Ninety to ninety-eight percent of potassium (K) found in the soil is tied up in primary minerals, thus making only 1 to 2% available in solution and on the exchange sites for plant uptake (Havlin et al., 2005). Deficiency of K shows up first in the older leaves and moves from older to younger leaves as the plant grows, since K is very mobile in the plant (Coombe, 1992). Potassium comprises 50 to 70% of the cations found in grape must and influences grape pH (an important variable in wine making). An excess of K in the grapes may cause high pH in wine, an unfavorable condition for the winemaker (Robinson, 1997).

Potassium deficiencies are easily confused with other nutrient deficiencies, so monitoring the vines by annual petiole testing is advised. Younger leaves can become shiny, show interveinal chlorosis, and have pronounced leaf blackening, uneven ripening, small and tight bunches. In white grape varieties, yellowing near the margin extending inwards towards the center of the leaf and can end up in marginal burning of the leaves (Coombe, 1992). In red grape varieties, the same occurs except with reddening instead of yellowing (Coombe, 1992).

Calcium (Ca)

Plants utilize calcium (Ca) in cell division and formation, making it important for overall root and vegetative plant growth (CFA, 1998). Calcium deficiencies are uncommon in California vineyards (Winkler et al., 1974). Soil Ca concentrations vary widely due to variables such as pH and geology (Havlin et al., 2005). The weathering of rocks like limestone and minerals such as plagioclase, calcite, and dolomite contribute a significant amount of Ca to the soil (Mottana et al., 1978). As soil pH rises, typically soil Ca levels increase (see Figure D1). Therefore, Ca is often most deficient in the highly acidic soils (pH < 6.0). Calcium deficiency symptoms include leaf discoloration, fruit disorders, and the premature abscission of buds and blossoms.

Table D1: Major Essential Mineral Nutrients of Grapevines, Mechanism of their Movement in Soils, and Effects of Nutrient Disorders

Mineral Nutrient	Mineral Nutrient Type	Principal Soil Movement Mechanism	Effects of Deficiency	Effects of Excess
N	Primary Macronutrient	Mass Flow and Diffusion	Reduced igor, small shoots and leaves, pale foliage, reduced yield	Excess vigor, enlarged leaves, reduced bud fruitfulness, reduced fruit set, reduced root growth
P	Primary Macronutrient	Diffusion	Retarded growth, reduced bud fruitfulness, reduced yield restricted foliage growth, reddened or yellow tissue between leaf veins	Zinc deficiency
K	Primary Macronutrient	Diffusion and Mass Flow	Leaf chlorosis and death, early leaf fall, retarded shoot growth, reduced cluster size and number, uneven ripening, increased susceptibility to frost damage	High fruit pH, decreased magnesium uptake
Ca	Secondary Macronutrient	Mass Flow and Root Interception	Restricted shoot and root growth	Reduced potassium and magnesium uptake
Mg	Secondary Macronutrient	Mass Flow and Root Interception	Yellowing at leaf edges, sometimes extending inward between main veins	Reduced potassium uptake
S	Secondary Macronutrient	Mass Flow	Uniform chlorosis, retarded plant growth, nitrogen accumulation	Soil acidification
B	Micronutrient	Mass Flow	Very short internodes, mottled and patched chlorosis, poor or no fruit set, shot berries, shoot tip death	Toxicity: Dark brown speckles or necrosis on edges of older leaves, cupped and wrinkled young leaves
Zn	Micronutrient	Mass Flow	Distorted, mottled apical leaves; stunted shoots; poor fruit set and shoot berries	Inhibited root growth, young root chlorosis
Mn	Micronutrient	Mass Flow	Chlorosis bands on basal leaves and death, decreased cold hardiness	Tissue injury, deficiency symptoms of other nutrients
Fe	Micronutrient	Mass Flow	Interveinal, creamy chlorosis on apical leaves; stunted shoots; reduced yield	Reduced yield
Cu	Micronutrient	Mass Flow	Short internodes, pale color, distorted young leaves	Reduced vigor, inhibited growth or root damage

Footnote: Figure D1 adapted from Grant, 2002.

Magnesium (Mg)

Plants utilize magnesium (Mg) for enzyme activation and as a main component of chlorophyll synthesis (CFA. 1995). Weathered rocks and minerals such as serpentine, biotite and dolomite supply magnesium to the soil (Havlin et al., 2005). In temperate soils, the Mg concentration in the soil solution ranges from 5 to 50 ppm (Tisdale, et al. 1985). Magnesium may also influence the pH of grape berries (Ruhl et al., 1992). Magnesium requirements vary among plants. Deficiencies are rare in California vineyards, except in those with sandy soils, calcareous (high Ca) soils, or soils with very high K levels (Winkler et al., 1974). Symptoms of mild magnesium deficiency have been seen in vineyards in the Mallee areas of Australia where calcareous soil are present (Coombe, 1992). Deficiency symptoms include the upward curling and red color of leaves and broad band chlorosis between the main veins (CFA, 1998; Coombe, 1992; see Photo D2). Mg deficiency can often be confused with zinc or iron deficiencies (Coombe, 1992).

Sulfur (S)

Sulfur (S) functions as a constituent of plant proteins and amino acids (CFA, 1998). Vineyards rarely suffer from an insufficient sulfur (S) supply (Winkler et al., 1974). Plants absorb a significant amount of S from "industrial enriched" air and precipitation. Sulfur is often added to vineyards during the use of S dust to combat powdery mildew (Winkler et al., 1974). Deficiency symptoms include retarded plant growth, delayed maturity, and the discoloration of younger leaves (Havlin et al., 2005).

Iron (Fe)

Plants utilize iron (Fe) in chlorophyll formation and biochemical processes such as respiration and photosynthesis (Havlin et al., 2005). Iron (Fe) deficiencies are rare in California vineyards with the exception of soils that have high manganese levels, calcareous soils (with greater than 5% calcium carbonate), or soils with poor aeration (Winkler et al., 1974; see Photo D1). Primary minerals and rocks such as basalt, gabbro, olivine and pyrite supply Fe to the soil (Havlin et al., 2005). Grapes are sensitive to low levels of available Fe in the soil solution (Havlin et al., 2005). Interveinal chlorosis of young leaves and twig dieback are common deficiency symptoms (CFA, 1998; see Photo D1). If interveinal chlorosis of leaves is seen in earlier stages of rootstock growth a foliar spray is recommended. A foliar spray using either iron chelate or iron sulfate may be applied directly to the vines. If an acidifying nitrogen fertilizer such as ammonium polyphosphate is applied, lower pH levels should allow for increased availability of the metal micronutrients (Beaton 1993).

Manganese (Mn)

Plants utilize manganese (Mn) as an enzyme activator for growth processes and in chlorophyll synthesis (CFA, 1998). Manganese (Mn) deficiencies appear to a "mild degree" in Californian vineyards (Winkler et al, 1974). Manganese can be found in ferromagnesian (Fe-Mg) rocks and secondary minerals such as rhodonite and manganite (Mottana et al., 1978). It also occurs in the Mn hydroxides and oxides that coat soil particles (Tisdale et al, 1985).

Copper (Cu)

Copper (Cu) is an enzyme activator and essential to vitamin A and protein production in plants (CFA, 1998). It is found in organic matter, and rocks and minerals, such as calcareous sedimentary rocks and chalocopyrite (Havlin et al., 2005). Copper deficiency has been seen in Gingin, Western Australia (Coombe, 1992). Vines were found to be have short canes with shortened internodes, the leaves were small, and had only slight indentations and were pale in color (Coombe, 1992). The excessive use of Cu fungicides can be toxic to plants. Toxicity symptoms include reduced shoot vigor, chlorosis, and poorly developed roots (Havlin et al., 2005).

Zinc (Fe)

Zinc (Zn) is a constituent of enzymes required for plant growth and is essential to the synthesis of indole acetic acid (IAA), a plant auxin (CFA, 1998). Sources of Zn include the minerals in many igneous and sedimentary rocks (Havlin et al., 2005). Zinc is the second most deficient metabolite in California vineyards (Winkler et al., 1974). Grapes are "very sensitive" to low levels of Zn (Havlin et al., 2005). For normal growth, plants require only about 0.5 lbs Zn per acre. Low levels are commonly found in soils that contain high P concentrations, low organic matter levels, poor aeration, or alkaline pH (Havlin et al., 2005). Grapevines grafted on nematode resistant rootstock may be susceptible to Zn deficiencies (Winkler et al., 1974). A severe Zn deficiency results in "little leaves" with mottling between the leaf veins, reduced berry size, and reduced grape numbers per cluster (Mullins et al., 1992; Coombe, 1992). The reduction of IAA production can inhibit bud formation and terminal growth (CFA, 1998).

Boron (B)

Boron (B) is also required for the elongation of pollen tubes and for "proper" cell division in various plant parts (Mullins et al., 1992). Plants utilize B to regulate carbohydrate metabolism and transport sugars (Burt et al., 1998). Sources of B include igneous rocks,

Photo D1: Lime-induced chlorosis indicating iron (Fe) and zinc (Zn) deficiencies in the calcareous Linne and Calodo soils located within the HMR Vineyards, located north of Peachy Canyon Road (2 June 2002).

Photo D2: Red-spotted Zinfandel leaves indicating magnesium (Mg) deficiency in these calcium (Ca)-rich and Mg-poor, calcareous Sorrento soils located in alluvial valleys at the Halter Ranch Vineyard (30 August 2003).

shale, clay minerals, and irrigation water (Havlin et al., 2005). A narrow window exists between the deficient and toxic levels of B in soil (Burt et al., 1998). Grapes have higher B requirements than most other deciduous fruit crops (Winkler et al., 1974). Deficiencies occur in high rainfall areas or irrigated sandy soils. Deficiency symptoms observed in the Granite Belt of Queensland, Australia included reduced set, high proportion of seedless berries, death of the shoot tip and yellowing between the veins of recently matured leaves (Coombe, 1992). Grapes are sensitive to levels of 0.5 to 0.75 ppm of B in overhead irrigation water (Havlin et al., 2005).

MOLYBDENUM (Mo)

Plants require molybdenum (Mo) for the utilization of N (CFA, 1998). California vineyards rarely display Mo deficiencies (Winkler et al., 1974). Molybdenum is found as a trace element in many primary and secondary minerals (Havlin et al., 2005). High levels of P may increase Mo concentrations. Soils, which are acidic, sandy, or those soils containing high levels of Zn, Cu, and Fe tend to induce Mo deficiency (Havlin et al., 2005). Plants lacking a sufficient supply of Mo may display stunted growth, marginal scorching of leaves, and a lack of vigor (CFA, 1998). Molybdenum toxicity can occur in alkaline pH soils (Havlin et al., 2005).

CHLORIDE (Cl)

Chloride (Cl) is involved in the light reaction of photosynthesis (Burt et al., 1998). Plants obtain it from weathered igneous and metamorphic rocks, sea air, precipitation, and irrigation water (Havlin et al., 2005). Leaf chlorosis followed by the wilting of the plant is a common deficiency symptom (CFA, 1998). High amounts of Cl can "burn" leaf tissue in grapes and create water stress for most plants by increasing the osmotic pressure of the soil solution (Winkler et al., 1974). There will likely be adequate Cl in the irrigation water to supply the grape vine needs.

SODIUM (Na; NOT A PLANT ESSENTIAL NUTRIENT)

Most plants do not require sodium (Na) for growth and avoid absorbing it (Burt et al., 1998). It is supplied to the soil through irrigation water, silicate minerals, and sodic plagioclases, like albite (Havlin et al., 2005). Its presence in soils concerns vineyard managers due its harmful effects (Winkler et al., 1974). High Na concentrations in the soil solution disperse organic and clay particles, which destroys soil permeability. It can also cause leaf burn and stunt vine growth (also common symptoms of a K deficiency). Five to ten meq/L (milliequivalents per liter) Na in irrigation water can cause injury to grapes (CFA, 1998). In many cases, gypsum ($CaSO_4 \cdot 2H_2O$) additions and leaching with good quality irrigation water are used to correct Na problems in agricultural soils (Tisdale et al., 1995).

ADDITIONAL INFORMATION SOURCES FOR VINEYARD MANAGEMENT

It is beyond the scope of this book to provide a comprehensive review of all the aspects of vineyard management related to soils. The bibliography should be consulted to obtain additional references on vineyard management related to soil fertility and fertilizer management, irrigation systems, and rootstock choices. An excellent starting point to access viticulture and enology information online is the University of California web site, Integrated Viticulture Online, at internet address: http://iv.ucdavis.edu (Univ. of California, 2007).

BIBLIOGRAPHY

Amerine, M.A. and A.J. Winkler. 1944. Growing degree-days. Hilgardia.

Arc Map. 2006. Release 9.2. ESRI Inc. Redlands, CA

Bailey, E.G., W.P. Irwin, and D.L. Jones. 1964. Franciscan and related rocks, and their significance in the geology of western California. Calif. Div. of Mines and Geology (now California Geological Survey) Bulletin 183, 177 pages.

Barbeau, D.L., Jr. Ducea, M.N., Gehrels, G.E., Kidder, S., Wetmore, P.H., and J. B. Saleeby. 2005. U-Pb detrital-zircon geochronology of northern Salinian basement and cover rocks. U.S. Geol. Soc. Amer. Bulletin 117(3-4):466-481.

Bates, Robert L. and J.A. Jackson (ed.). 1984. Dictionary of geologic terms. Doubleday, New York, N.Y.

Baumgartner, K. 2003. Encouraging beneficial AM fungi in vineyard soil. Practical Winery & Vineyard. January/February 2003.

Bauman, A. and Gabriel, T. 1997. Detailed Soil Survey of the Windward Vineyard, Templeton Gap, San Luis Obispo County, California. T.J. Rice, Advisor. Cal Poly Senior Project, Earth and Soil Sciences Dept., San Luis Obispo, CA. Prepared for Windward Vineyard, 1380 Live Oak Road, Paso Robles, CA.

Beardsley, T. 1988. *In Vino Scientia*: Science is improving California's premium wines. Scientific Amer. 259:41.

Beck, C., J. Grieser, M. Kottek, F. Rubel, and B. Rudolf. 2006. Characterizing Global Climate Change by means of Köppen Climate Classification. *Klimastatusbericht*, 2005:139-149.

Berry, E. 1990. The importance of soil in fine wine production. Journal of Wine Research 1:179-194.

Bettiga, L. et al. 2003. Wine grape varieties in California. Publ. 3419, Univ. of Cal Agric. and Natural Resources. Oakland, CA.

Blakey, R.C. 2007. Paleogeography through geologic time. Internet Source: http://jan.ucc.nau.edu/~rcb7/global_history.html (Verified: 20 August 2007).

Bohmrich, R. 1996-7. *Terroir.* Journal of Wine Research 7(1): 33-46.

Boulton, R. 1980. A hypothesis for the presence, activity, and role of potassium/hydrogen, adenosine triphosphatases in grapevines. American Journal of Enology and Viticulture 31(3):283-287.

Boulton, R. 2001. Titratable acidity and pH – their importance and their adjustment. Presented at the 16th Midwest Regional Grape and Wine Conference, February 4, 2001, Osage Beach, Missouri.

Brady, N.C. and R.R. Weil. 2004. Elements of the Nature and Properties of Soils. Second edition. Pearson Education Inc., Upper Saddle River, NJ.

Burt, C. K. O'Connor and T. Ruehr. 1998. Fertigation. Irrigation Training and Research Center, California Polytechnic State Univ., San Luis Obispo, CA.

Burt, Charles. 1998. Irrigation water management. BioResource and Agricultural Engineering Dept. California Polytechnic State Univ., San Luis Obispo, CA.

California Fertilizer Association (CFA). 1998. Western Fertilizer Handbook. Interstate Publ., Inc. Danville, IL.

Carbonneau, A.P. and P. Casteran. 1987. Interactions "training system x soil x rootstock" with regard to vine ecophysiology, vigour, yield and red wine quality in the Bordeaux area. *Acta Horticulturae* 206:119-140.

Carbonneau, A.P. 1985. The early selection of grapevine rootstocks for resistance to drought conditions. American Journal of Enology and Viticulture, 36:195-198.

Central Coast Vineyard Team. 1998. Positive Points System. Practical Winery and Vineyard, May/June 1998.

Chen, Y. and P. Barak. 1982. Iron nutrition of plants in calcareous soils. Advances in Agronomy 35:217-240.

Chiaramonte, M. 2002. Water holding capacity of rock types in the Monterey Formation, Sandholdt member. SS 400 project report, T.J. Rice, faculty dvisor. Cal Poly, San Luis Obispo, CA.

Chipping, D.H. 1987. The geology of San Luis Obispo Co., CA; A brief description and field guide. El Corral, Cal Poly, San Luis Obispo, CA.

Chone, X., C. Van Leeuwan, P. Chery, and P. Ribereau. 2001. *Terroir* influence on water status of non-irrigated Cabernet Sauvignon (Vitus vinifera): vegetative development, must and wine composition. South African Journal of Enology and Viticulture 22(1): 8-15.

Christensen, L.P. 2003. Rootstock selection, p. 12-15. *In* Wine Grape Varieties in California. Univ. of California Agricultural and Natural Resources Publication 3419, Oakland, CA.

Christensen, L.P. 2000. Use of tissue analysis in viticulture. Grape Notes. Sept.-Oct. 2000 UC Coop. Ext. Newsletter. Tulare, CA.

Christensen, L.P. 1990. How much nitrogen does a vine use? Hot News In Viticulture. Kearney Agriculture Center Publication, Univ. Calif., Parlier, CA.

Christensen, L.P., A.N. Kasamatis, and F.L. Jensen. 1978. Grapevine nutrition and fertilization in the San Joaquin Valley. Publication 4087, Division of Agricultural Sciences. Univ. of California. Oakland, CA. ISBN 0-931876-25-7.

Climatological Data. 2007a. National Climate Data Center, U.S. National Oceanic and Atmospheric Administration (NOAA). Internet Source: http://www.ncdc.noaa.gov/oa/ncdc.html/

Climatological Data. 2007b. Western Regional Climate Center, Desert Research Institute, Univ. of Nevada, Reno. Internet Source: http://www.wrcc.dri.edu/

Coombe, B.G. and P.R. Day (eds.). 1992a. Viticulture: Vol. 1, Resources. Winetitles, Adelaide, Australia

Coombe, B.G. and P.R. Day (eds.). 1992b. Viticulture: Vol. 2, Practices. Winetitles, Adelaide, Australia

Cui, Z., Tjernstrom, M. and B. Grisogono. 1998. Idealized simulations of atmospheric coastal flow along the central coast of California. Journal of Applied Meteorology, 37: 1332-1345.

Delas, J. and R. Pouget. 1979. The influence of rootstock-scion relationships on the mineral nutrition of the vine and the consequences for fertilization. p. 289-300. *In* Soils in Mediterranean type climates and their yield potential. Intl. Potash Inst., Bern.

Dibblee, T.W., Jr. 2007. Geology map of the Lime Mountain Quadrangle, San Luis Obispo County, CA. 1:24,000 scale. First printing, Feb., 2007. Dibblee Center Map #DF-285. Ian Campbell Memorial Map. Santa Barbara Museum of Natural History.

Dibblee, T.W., Jr. 2006a. Geology map of the Adelaida Quadrangle, San Luis Obispo County, CA. 1:24,000 scale. First printing, May 2006. Dibblee Center Map #DF-218. David L. Durham Honorary Map. Santa Barbara Museum of Natural History.

Dibblee, T.W., Jr. 2006b. Geology map of the San Miguel Quadrangle, Monterey and San Luis Obispo Counties, CA. 1:24,000 scale. First printing, May 2006. Dibblee Center Map #DF-220. Santa Barbara Museum of Natural History.

Dibblee, T.W., Jr. 2004a. Geology map of the Atascadero Quadrangle, San Luis Obispo County, CA. First printing, June 2004. Dibblee Center Map #DF-132. Ed Hart Honorary Map. Santa Barbara Museum of Natural History.

Dibblee, T.W., Jr. 2004b. Geology map of the Creston & Shedd Canyon Quadrangles, San Luis Obispo County, CA. First printing, June 2004. Dibblee Center Map #DF-136. Santa Barbara Museum of Natural History.

Dibblee, T.W., Jr. 2004c. Geology map of the Estrella & Shandon Quadrangles, San Luis Obispo County, CA. First printing, June 2004. Dibblee Center Map #DF-138. Santa Barbara Museum of Natural History.

Dibblee, T.W., Jr. 2004d. Geology map of the Lopez Mountain Quadrangle, San Luis Obispo County, CA. First printing, June 2004. Dibblee Center Map #DF-130. Hugh McLean Honorary Map. Santa Barbara Museum of Natural History.

Dibblee, T.W., Jr. 2004e. Geology map of the Paso Robles Quadrangle, San Luis Obispo County, CA. First printing, June 2004. Dibblee Center Map #DF-137. Santa Barbara Museum of Natural History.

Dibblee, T.W., Jr. 2004f. Geology map of the San Luis Obispo Quadrangle, San Luis Obispo County, CA. First printing, June 2004. Dibblee Center Map #DF-129. Frank Dennison Honorary Map. Santa Barbara Museum of Natural History.

Dibblee, T.W., Jr. 2004g. Geology map of the Santa Margarita Quadrangle, San Luis Obispo County, CA. First printing, June 2004. Dibblee Center Map #DF-133. Santa Barbara Museum of Natural History.

Dibblee, T.W., Jr. 2004h. Geology map of the Templeton Quadrangle, San Luis Obispo County, CA. First printing, June 2004. Dibblee Center Map #DF-135. Lew Rosenberg Honorary Map. Santa Barbara Museum of Natural History.

Dibblee, T.W., Jr. 1979. Cenozoic tectonics of the northwest flank of the Santa Lucia Mountains from the Arroyo Seco to the Nacimiento River, CA, p. 67-76. In S.A. Graham (ed.). Tertiary and Quaternary geology of the Salinas Valley and Santa Lucia Range, Monterey Co., CA. Pacific Coast Paleogeography Field Guide 4.

Dibblee, T.W., Jr. 1976. The Rinconada and related faults in the southern Coast Ranges, CA and their tectonic significance. U.S. Geolgoical Survey Prof. Paper 981, 55 pages

Dibblee, T.W., Jr. 2004i. Geology of the San Luis Obispo quadrangle, CA. U.S. Geological Survey Open File Report 74-224, 1:62,500 scale.

Dibblee, T.W., Jr. 1974. Geologic map of the Shandon and Orchard Peak quadrangles, San Luis Obispo and Kern Counties, CA, showing Mesozoic and Cenozoic rock units juxtaposed along the San Andreas Fault. U.S. Geological Survey Misc. Investigations Series Map I-788, 1:62,500 scale.

Dibblee, T.W., Jr. 1971. Geology maps of seventeen 15-minute quadrangles along the San Andreas fault in the vicinity of King City, Coalinga, Panoche Valley and Paso Robles, CA; with index map. U.S. Geological Survey Open File Maps, 1:62,500 scale.

Dickenson, J. 1990. Viticultural geography: an introduction to the literature in English. Journal of Wine Research 1: 5-24.

Doyle, J.D. 1997. The influence of mesoscale orography on a coastal jet and rainband. Monthly Weather Review 125:1465-1488.

Ducea, M., M.A. House and S. Kidder. 2003. Late Cenozoic denudation and uplift rates in the Santa Lucia Mountains, California. Geology, 31(2):139-142.

Durham, D.L. 1974. Geology of the southern Salinas Valley area, CA. U. S. Geological Survey Professional Paper 819, 106 pages.

Durham, D.L. 1968. Geologic map of the Adelaida Quadrangle, San Luis Obispo County, CA. U. S. Geol. Survey Map GQ-768.

Elliott-Fisk, D. 2007. Geographical and viticultural diversity of the Paso Robles AVA. Included as Appendix C in Comment #98 from the Paso Robles AVA Committee to TTB Notice No. 71, Proposed Establishment of the Paso Robles Westside Viticultural Area. Internet Source: http://www.ttb.gov/nprm_comments/ttbnotice71_comments.shtml.

Eswaran, H., T.J. Rice, B.A. Ahrens and B.L. Stewart (ed.). 2003. Soil Classification: A World Desk Reference. CRC Press.
Fregoni, M. 1977. Effects of soil and water on the harvest. Proceedings of the International Symposium on the Quality of the Vintage. Cape Town, South Africa.

Finch, C.U. and W.C. Sharp. 1983. Cover Crops in California Orchards & Vineyards. Soil Conservation Service (now Natural Resources Conservation Service), Davis, CA.

Fugro West, Inc. and Cleath & Associates. 2002. Final Report Paso Robles Groundwater Basin Study, Phase I. Internet Source: http://www.slocountywater.org/site/Water Resources/, Phase 1.Reports/Paso Phase 1/index.htm

Fugro West, Inc., ETIC Engineering, Inc., and Cleath & Associates. 2005. Final Report Paso Robles Groundwater Basin Study, Phase II. Numerical Model Development, Calibration and Application. Internet Source: http://www.slocountywater.org/site/Water Resources/, Phase 1.Reports/Paso Phase 2/index.htm

Gilbride, B. and Van Trump, G. 1999. Detailed Soil Survey of the Valino Vineyard, Creston, San Luis Obispo County, California. T.J. Rice, Advisor. Cal Poly Senior Project, Earth and Soil Sciences, San Luis Obispo, CA.

Grant, R.S. 2002. Balanced soil fertility management in wine grape vineyards. Practical Winery & Vineyard, May/June 2002.

Grant, R.S. 1999. Managing phosphorus deficiency in vineyards. Practical Winery & Vineyard, January/February 1999.

Greenlough, J.D., H.P. Longerich, S.E. Jackson. 1997. Element fingerprinting of Okanagan Valley wines using ICP-MS: Relationships between wine composition, vineyard and wine colour. Australian Journal of Grape and Wine Research 3:75-83.

Hall, C.A. and S.W. Prior. 1973. Geology of the Cayucos-San Luis Obispo region, San Luis Obispo County, CA. U.S. Geological Survey Misc. Field Studenies Map MF-680. 1:48,000 scale.

Halliday, J. and H. Johnson. 1992. The Art and Science of Wine. London.

Hardie, W.J. and R.M. Cirami. 1992. Grapevine rootstocks, p. 154-176. *In* B.G. Coombe and P.R. Day (eds.). Viticulture: Vol. 1, Resources. Winetitles, Adelaide, Australia.

Hart, E.W. 1976. Basic geology of the Santa Margarita area, San Luis Obispo County, CA. Calif. Div. of Mines and Geol. Bulletin 199. CA Geological Survey, Sacramento, CA.

Hart, M. 1999. Detailed Soil Survey of the future Bartholomew Vineyard, Santa Rosa Creek Road, San Luis Obispo county, California. T.J. Rice, Advisor. Cal Poly Senior Project, Earth and Soil Sciences Dept., San Luis Obispo, CA.

Havlin, J.L., S.L. Tisdale, J.D. Beaton, and W.L. Nelson. 2005. Soil Fertility and Fertilizers. Seventh edition. Pearson Education, Inc., Upper Saddle River, NJ.

Hornbeck, D. 1983. California Patterns: A Geographical and Historical Atlas. Univ. of California Press, Berkeley.

Jennings, C. W. 1958. San Luis Obispo Sheet, Geologic Map of California Series, Division of Mines and Geology (now California Geological Survey), Dept. of Conservation, The Resources Agency, State of California, Sacramento.

Jennings, C.W. 1977. Geologic Map of California. California Division of Mines and Geology (now California Geological Survey), Dept. of Conservation, The Resources Agency, State of California, Sacramento, CA.

Klein, C. and C. Hurlbut. 1993. Manual of Mineralogy. 21st ed. John Wiley & Sons, Inc., New York.

Koppen, G., Geiger and Pohl. 1953. Koppen-Geiger Climatic Classification of the World.

Kottek, M., J. Grieser, C. Beck, B. Rudolf, and F. Rubel. 2006. World Map of the Köppen-Geiger climate classification updated. Meteorol. Z., 15, 259-263.

Kuchler, A. W., 1977. Potential Natural Vegetation of California Map. Univ. of California Press, Berkeley.

LeClair, K.A. and Rebel, T.D. 1997. Chardonnay Grape Chemistry Related to Soil Chemistry at Carmody McKnight (Silver Canyon) Vineyards, Adelaida, California. T.J. Rice, Advisor. Cal Poly Senior Project, Earth and Soil Sciences, San Luis Obispo, CA.

Loveys, B., M. Stoll, P. Dry, and M. McCarthy. 1998. Partial rootzone drying stimulates stress responses in grapevine to improve water use efficiency while maintaining crop yield and quality. Australian Grapegrower and Winemaker, Annual Technical Issue, p. 108-113.

McGrath, M. 2001. Detailed Soil Survey for the future Crawford Vineyards, Creston, California. T.J. Rice, Advisor. Cal Poly Senior Project, Earth and Soil Sciences Dept., San Luis Obispo, CA

Meirik, E. 2005. Soil Resource Assessment of Selected Vineyards in the Paso Robles AVA, San Luis Obispo County, California. T.J. Rice, Advisor. Cal Poly Senior Project, Earth and Soil Sciences Dept., San Luis Obispo, CA

Merenlender, A.M. and J. Crawford. 1998. Vineyards in an Oak Landscape. Univ. of Calif. Div. of Agric. and Natural Resources Publication 21577. Oakland, CA.

Miller, R. and D. Gardiner. 1998. Soils in Our Environment. 8th ed. Prentice-Hall Inc., Englewood Hills, NJ.

Mitchell, C.C. 1999. Interpreting soil test results. Alabama Cooperative Extension System.

Moran, W. 2006. You said *Terroir*? Approaches, sciences and explanations. Keynote address, *Terroir* 2006 Conference, Univ. of Calif., Davis. 19 to 23 March 2006.

Mottana, A., R. Crespi, and G. Liborio. 1978. Guide to Rocks and Minerals. Simon and Schuster, Inc., New York, NY.

Mullins, M.G., A. Bouquet, and L.E. Williams. 1992. Biology of the Grapevine. Cambridge Univ. Press, Great Britain.

Nelson, R.E. 1982. Carbonate and gypsum, p. 181-197. *In* A.L. Page (ed.) Methods of soil analysis. Part 2. 2nd ed. Agron. Monogr. 9. ASA and SSSA, Madison, WI.

Nicholson, L. 1980. Rails Across the Ranchos. Valley Publishers, Fresno, CA.

Norris, R.M. and R.W. Webb. 1990. Geology of California. John Wiley & Sons, New York.

Ohles, W.V. 1997. The Lands of Mission San Miguel. World Dancer Press, Clovis, CA.

Page, B.M., G.A. Thompson and R.G. Coleman. 1998. Late Cenozoic tectonics of the central and southern Coast Ranges of California. Geol. Soc. Amer. Bulletin, 110(7):846-876.

Paso Robles Wine County Alliance. Distinctly Different Paso Robles. Accessed May 2007. Internet Source: http://www.pasowine.com/pasorobles/history.php

Peacock, B. 1995. Water management for grapevines. Publ. #IG1-95. (newsletter) Univ. of California Cooperative Extension, Tulare Co., Tulare, CA.

Peacock, B. and L.P. Christensen. 1996. Potassium and boron fertilization in vineyards. Publ. #NG1-96. (newsletter) Univ. of California Cooperative Extension, Tulare Co., Tulare, CA. Internet Source: http://cetulare.ucdavis.edu/pubgrape/ng196.htm

Peacock, B. and D. Handley. 1998. Drip irrigation must apply water uniformly to be efficient. Publ. #IG2-95. (newsletter) Univ. of California Cooperative Extension, Tulare Co., Tulare, CA.

Peyrout des Gachons, C., C. Van Leeuwan, C. Tominga, J.-P. Gaudillere and D. Dubourdieu. 2005. The influence of water and nitrogen deficit on fruit ripening and aroma potential of *Vitus vinifera* L. cv Sauvignon Blanc in field conditions. Journal of the Science of Food and Agriculture 85(1):73-85.

Pouget, R. 1987. Usefulness of rootstocks for controlling vine vigour and improving wine quality. *Acta Horticulturae* 206:109-119.

Renner, R. 1990. Does the secret lie in the soil? New Scientist 22:62-64.

Reynier, A. 1997. Manuel de viticulture (in French). Lavoisier Publ., Paris.

Rice, T.J. 2006. Soil Resource Inventory. El Corral, San Luis Obispo, CA.

Rice, T.J. 2002a. Importance of soil texture to vineyard management. Practical Winery and Vineyard magazine, March/April 2002. San Rafael, CA.

Rice, T.J. 2002b. Soil survey report for the Halter Ranch Vineyard, Paso Robles, California. Prepared for Halter Ranch Vineyard, 8910 Adelaida Road, Paso Robles, CA.

Rice, T.J. 2001. Soil survey report for future China Cove Vineyards, Cayucos, California. Prepared for Kidd Vineyard Development, San Luis Obispo, CA.

Rice, T.J. 2000. Soil survey report for the future Halley Vineyards, Paso Robles, California. Prepared for Bob Halley, Ambush Trail, Paso Robles, CA.

Rice, T.J. 1999a. Liming of vineyard soils. Practical Winery and Vineyard magazine, July/August 1999. San Rafael, CA.

Rice, T.J. 1999b. Soil survey report for the Preston Vineyards, Templeton, California. Prepared for Jack and Charlotte Preston, Cobble Creek Road, Templeton, CA.

Rice, T.J. 1997a. A comprehensive vineyard study for Carmody McKnight and Justin vineyards. *In* G. Conway (ed.). KCBX Central Coast Wine Classic vineyard symposium guidebook, July 11, 1997. Conway and Justin Vineyards, Adelaida Hills, CA.

Rice, T.J. 1997b. Soil survey report for the DeBro Vineyards, Adelaida, California. Prepared for Richard and Sheri DeBro, 10950 Chimney Rock Road, Paso Robles, CA.

Rice, T.J and Cal Poly, Soil Resource Inventory (SS 431) students. 2007. Soil survey report for the L'Aventure Winery and Stephan Vineyard, Paso Robles, California. Prepared for L'Aventure Winery and Stephan Vineyard, 2815 Live Oak Road, Paso Robles, CA.

Rice, T.J and Cal Poly, Soil Resource Inventory (SS 431) students. 2006. Soil survey report for the Halter Ranch Organic Farm, Paso Robles, California. Prepared for Halter Ranch Vineyard, 8910 Adelaida Road, Paso Robles, CA.

Rice, T.J and Cal Poly, Soil Resource Inventory (SS 431) students. 2005. Soil survey report for the Halter Ranch West, Paso Robles, California. Prepared for Halter Ranch Vineyard, 8910 Adelaida Road, Paso Robles, CA.

Rice, T.J and Cal Poly, Soil Resource Inventory (SS 431) students. 2004. Soil survey report for the Halter Ranch East (east part), Paso Robles, California. Prepared for Halter Ranch Vineyard, 8910 Adelaida Road, Paso Robles, CA.

Rice, T.J and Cal Poly, Soil Resource Inventory (SS 431) students. 2003. Soil survey report for the Halter Ranch East (west part), Paso Robles, California. Prepared for Halter Ranch Vineyard, 8910 Adelaida Road, Paso Robles, CA.

Rice, T.J and Cal Poly, Soil Resource Inventory (SS 431) students. 2002a. Soil survey report for the South Viking future vineyards, east portion, Peachy Canyon Road, Paso Robles, CA. Prepared for Adelaida Cellars, 5805 Adelaida Road, Paso Robles, CA.

Rice, T.J and Cal Poly, Soil Resource Inventory (SS 431) students. 2002b. Soil survey report for the South Viking future vineyards, west portion, Peachy Canyon Road, Paso Robles, CA. Prepared for Adelaida Cellars, 5805 Adelaida Road, Paso Robles, CA.

Rice, T.J and Cal Poly, Soil Resource Inventory (SS 431) students. 2001a. Soil survey report for the HMR Vineyards, west portion, Peachy Canyon Road, Paso Robles, CA. Prepared for Adelaida Cellars, 5805 Adelaida Road, Paso Robles, CA.

Rice, T.J and Cal Poly, Soil Resource Inventory (SS 431) students. 2001b. Soil survey report for the HMR Vineyards, undeveloped northern lands portion, Peachy Canyon Road, Paso Robles, CA. Prepared for Adelaida Cellars, 5805 Adelaida Road, Paso Robles, CA.

Rice, T.J and Cal Poly, Soil Resource Inventory (SS 431) students. 2000. Soil survey report for the HMR Vineyards, central portion, Peachy Canyon Road, Paso Robles, CA. Prepared for Adelaida Cellars, 5805 Adelaida Road, Paso Robles, CA.

Rice, T.J and Cal Poly, Soil Resource Inventory (SS 431) students. 1999. Soil survey report for the HMR Vineyards, east portion, Peachy Canyon Road, Paso Robles, CA. Prepared for Adelaida Cellars, 5805 Adelaida Road, Paso Robles, CA.

Rice, T.J. and K.P. Patterson. 2000. Soil survey report for the future Orcutt Hills Vineyards, Orcutt, California. Prepared for Nuevo Energy Company, Houston, Texas.

Roberts, D.R., and C.P. Stubler. 1996. Chemical analysis of the soils of the Carmody McKnight (Silver Canyon) Vineyard. Cal. Poly. Senior Project. Soil Sci. Dept., San Luis Obispo, CA.

Robinson, J. (ed.). 2006. The Oxford Companion to Wine. Oxford Univ. Press, New York, NY.

Robinson, J.B. 1992. Grapevine Nutrition. p. 178-208. In B.G. Coombe and P.R. Day (eds.). Viticulture: Vol. 2, Practices. Winetitles, Adelaide, Australia.

Ruhl, E.H., A.P. Fuda, and M.T. Treeby. 1992. Effect of potassium, magnesium, and nitrogen supply on grape juice composition of Riesling, Chardonnay, and Cabernet Sauvignon vines. Australian J. Expt. Ag. 32:645-649.

Schoeneberger, P.J., Wysocki, D.A., Benham, E.C. and Broderson, W.D. (ed.). 2002. USDA-NRCS, Field book for describing and sampling soils, version 2.0. Natural Resources Conservation Service, National Soil Survey Center, Lincoln, NE.

Scott, C.E. 1996. Bedrock and Soil Chemistry of the Tablas Creek Vineyards, Adelaida, California. T.J. Rice, Advisor. Cal Poly Senior Project, Earth and Soil Sciences, San Luis Obispo, CA.

Seiders, V.M. 1982. Geologic map of an area near York Mountain, San Luis Obispo County, CA. U.S. Geological Survey Misc. Investigations Series Map I-1369, 1:24,000 scale.

Sequin, G. 1986. "Terroirs" and pedology of wine growing. Experientia 42:861-872.

Soil Survey Staff, 2007. Official soil series descriptions. U.S. Dept. of Agric., Natural Resources Conservation Service. Accessed May, 2007. Internet Source: http://soils.usda.gov/technical/classification/osd/index.html

Silverman, M. and S. Edblad, 2001. Detailed Soil Survey of the Future HMR Vineyards, western part, Adelaida, California. T.J. Rice, Advisor. Cal Poly Senior Project, Earth and Soil Sciences, San Luis Obispo, CA. Prepared for Adelaida Cellars, 5805 Adelaida Road, Paso Robles, CA.

Smart, R.E. 1997. Partial root zone drying: vineyard irrigation breakthrough? Practical Winery & Vineyard March/April 1997.

Smart, R.E. 1996. Vineyard design to improve wine quality in the Orlando way. Australian and New Zealand Wine Industry Journal 11:335-336.

Smart, R.E. 1991. Sunlight into Wine: A Handbook for Winegrape Canopy Management. Winetitles, Underdale, South Australia.

Soil Science Society of America (SSSA). 2007. Glossary of Soil Science Terms. Internet Source: "http://www.soils.org/sssagloss/"

Soil Survey Staff. 2006. Keys to soil taxonomy, 10th edition. Natural Resources Conservation Service, U.S. Department of Agriculture. U.S. Govt. Printing Office, Washington, D.C.

Soil Survey Staff. 1999. Soil taxonomy: A basic system of soil classification for making and interpreting soil surveys. 2nd edition. Natural Resources Conservation Service. U.S. Department of Agriculture Handbook 436. U.S. Govt. Printing Office, Washington, D.C.

Sloan, R., and W. Westerling. 1996. Soil survey report of the Carmody McKnight (Silver Canyon) Vineyard. T.J. Rice, Advisor. Cal. Poly. Senior Project, Soil Sci. Dept., San Luis Obispo, CA.

Soil Survey Staff. 1993. Soil Survey Manual. USDA Handbook No. 18. U.S. Govt. Print. Office, Washington, D.C.

Taliaferro, N.L. 1943. Geologic history and structure of the central Coast Ranges of California. Cal. Div. of Mines and Geology (now California Geological Survey) Bulletin 118, p. 119-163.

Thomas, G. 1996. Soil pH and soil acidity, p. 476-477. *In* D. L. Sparks (ed). Methods of soil analysis. Part 3. Chemical methods. Soil Science Society of America, Inc., Madison, WI.

Tisdale, S.L., W.L. Nelson, J.D. Beaton and J.L. Havlin. 1985. Soil Fertility and Fertilizers. Macmillan Publ. Co., New York, NY.

USDA-Natural Resources Conservation Service (USDA-NRCS). 2003. Keys to soil taxonomy. 9th edition. U.S. Department of Agriculture, Natural Resources Conservation Service (NRCS). U.S. Govt. Printing Office, Washington, D.C. (NOTE: The soils in the Paso Robles Area soil survey report were classified according to earlier editions of this publication.)

United States Department of Agriculture (USDA). 2003. Soil survey of San Luis Obispo county, California; Carrizo Plain Area. (E.N. Vinson and K. Oster, primary authors) U.S. Govt. Printing Office, Washington, D.C.

United States Department of Agriculture (USDA). 1995. Soil survey manual. USDA Handb. 18. U.S. Govt. Print. Office, Washington, D.C. Printing Office, Washington, D.C.

United States Department of Agriculture (USDA). 1983. Soil Survey of San Luis Obispo County; Paso Robles Area; California. (W.C. Lindsey, primary author) U.S. Govt. Print. Office, Washington, D.C.

United States Geological Survey (USGS). 2007. Farallon Plate, tectonic geologic history. Internet Source: http://pubs.usgs.gov/gip/dynamic/Farallon.html (verified 20 August 2007).

Univ. of California. 2007. Integrated Viticulture Online. Internet Source: http://iv.ucdavis.edu/

Valentine, D.W., J.N. Densmore, D.L. Galloway and F. Amelung. 2001. Use of InSAR to identify land surface displacements caused by aquifer-system compaction in the Paso Robles area, San Luis Obispo County, California, March to August 1997. U.S. Geological Survey, Open File Report 00-447, Menlo Park, CA.

Vierra, S. and M. Mayes.1999. Detailed Soil Survey of Laura's Vineyards, Paso Robles, San Luis Obispo County, California. T.J. Rice, Advisor. Cal Poly Senior Project, Earth and Soil Sciences Dept., San Luis Obispo, CA.

Walker, M.A. 2006. Rootstock selection. Wine grape syllabus. U.C. Coop. Extension Agriculture and Natural Resources. Davis, CA

Williams, L.E. and P.J. Biscay. 1991. Partitioning of dry weight, nitrogen, and potassium in Cabernet Sauvignon grapevines from anthesis until harvest. American Journal of Enology and Viticulture 42(2):113-117.

Winkler, A.J., J.A. Cook, W.M. Kliewer, and L.A. Lider. 1974. General Viticulture. 2nd ed. Univ. Calif. Press, Berkeley, CA.

Wolpert, J.A., M.A. Walker and E. Weber (editors). 1992. Proceeding of Rootstock Seminar: A Worldwide Perspective, Reno, Nevada, June 24, 1992. Publ. by the American Society of Enology & Viticulture.

Wolpert, J.A., D.R. Smart and M. Anderson. 2005. Lower petiole potassium concentration at bloom in rootstocks with *Vitis berlandieri* genetic backgrounds. American Journal of Enology and Viticulture 56(2): 163-169.

PHOTO CREDITS

All of the numbered and un-numbered photographs in this book are original images produced by Thomas J. Rice, Ph.D., using an Olympus 5050-D digital camera, except those listed below, according to book page numbers.

Anthony Bower (photographer):
1.) p. 47: Rabbit Ridge winery buildings.
2.) p. 48: Erich Russell.
3.) p. 48: Jim Gibbons on motorcycle.
4.) p. 49: Jim and Shirley Gibbons in tasting room.
5.) p. 52: Grape vine on Vista del Rey vineyard.
6.) p. 52: Dave King in tasting room.
7.) p. 97: Jerry Lohr in tractor.
8.) p. 98: Arciero winery entrance.

Supplied by wineries or downloaded from winery web pages:
1.) p. 51: Vista del Rey label, found on winery web page.
2.) p. 72: Treana logo and wine bottles, found on winery web page.
3.) p. 96: Gary Eberle and poodles, provided by Eberle Winery.
4.) p. 99: Garretson family, provided by Matt Garretson.
5.) p. 104: Robert Hall in vineyard, provided by Robert Hall.
6.) p. 106: Three photos provided by Cass Winery (harvest bins, harvest truck, & women stomping grapes).
7.) p. 105: Gelfand Vineyards logo, found on winery web page.
8.) p. 111: The Midlife Crisis Winery logo, found on winery web page.
9.) p. 117: McClean home, found on winery web page.
10.) p. 117: Victor Hugo Winery barn, provided by Vic Roberts.
11.) p. 121: Cayucos Cellars owners in tasting room, found on winery web page.
12.) p. 134: Carmody McKnight vineyards and Santa Lucias with snow cover, provided by Gary Conway.

APPENDIX C

Contact Information for Wineries in the Paso Robles AVA

5 Mile Bridge
Phone (P) 805-788-4588
Fax (F) 805-595-2407
URL www.5milewine.com
Email vance@5milewine.com

Adelaida Cellars
5805 Adelaida Road
Paso Robles, CA. 93446
P 805-239-8980
F 805-239-4671
URL www.adelaida.com
Email wines@adelaida.com

AJB Cellars
3280 Township Road
Paso Robles, CA. 93446
P 805-239-9432
F 310-379-1679
URL www.ajbvineyards.com
Email ajbvineyards@msn.com

Anglim Winery
740 Pine Street
Paso Robles, CA. 93446
P 805-227-6813
F 805-888-2717
URL www.anglimwinery.com
Email info@anglimwinery.com

Aron Hill Vineyards
2500 Vineyard Drive
Templeton, CA. 93465
Arroyo Robles Winery
739 12th Street
Paso Robles, CA. 93446
P 877-759-9463
URL www.arroyorobles.com
Email jason@arroyorobles.com

Asuncion Ridge Vineyards
11010 Fuentes Road
Atascadero, CA 93442
P 805-461-0675
URL www.ascuncionridge.com
Email ascucionridge@hughes.net

B &E Vineyard/Winery
10,000 Creston Road
Paso Robles, CA 93446
P 805-238-4815
F 805-237-0805
Email bevineyard@aol.com

Bella Luna Winery
1850 Templeton Road
Paso Robles, CA. 93465
P 805-434-5477
F 805-434-5479
URL www.bellalunawine.com

Bianchi Winery
3380 Branch Road
Paso Robles, CA. 93446
P 805-226-9922
F 805-226-8230
URL www.bianchiwine.com
Email tr@bianchiwine.com

Booker Winery
2640 Anderson Road
Paso Robles, CA. 93446
P 805-610-2272
F 805-237-7368
URL www.bookerwines.com
Email jensenvineyards@aol.com

Brian Benson Cellars
2985 Anderson Road
Paso Robles, CA. 93446
P 805-296-9463
F 805-226-9467
URL www.brianbensoncellars.com
Email brian@brianbensoncellars.com

Calcareous Vineyard
3430 Peachy Canyon Road
Paso Robles, CA. 93446
P 805-239-0289
F 805-239-0916
URL www.calcareous.com
Email service@calcareous.com

Caparone Winery
2280 San Marcos Road
P 805-467-3827
F 805-238-9416
URL www.caparone.com
Email info@caparone.com

Carmody McKnight Estate
11240 Chimney Rock Road
Paso Robles, CA. 93446
P 805-238-9392
F 805-238-3975
URL www.carmodymcknight.com
Email info@carmodymcknight.com

Casa De Caballos
2225 Raymond Ave.
Templeton, CA. 93465
P 805-434-1687
F 805-434-1560
URL www.casadecaballos.com
Email info@casadecaballos.com

Cass Winery
7350 Linne Road
Paso Robles, CA. 93446
P 805-239-1730

F 805-227-2889
URL www.casswines.com
Email info@casswines.com

Castoro Cellars
1315 N. Bethel Road
Templeton, CA. 93465
P 805-238-0275
F 805-238-2602
URL www.castorocellars.com
Email info@castorocellars.com

Cayucos Cellars
143 N. Ocean Ave.
Cayucos, CA. 93430
P 805-995-3036
F 805-995-2415
URL www.cayucoscellars.com
Email wine@cayucos.cellars.com

Chateau Margene
4385 La Panza Road
Creston, CA. 93432
P 805-238-2321
F 805-238-2118
URL www.chateaumargene.com
Email info@chateaumargene.com

Christian Lazo Wines
249 10th Street, Suite A
San Miguel, CA. 93451
P 805-467-2672
URL www.christianlazowines.com
Email orders@christianlazowines.com

Chumeia Vineyards
8331 Highway 46 East
Paso Robles, CA 93446
P 805-226-0102
F 805-226-0104
URL: www.chumeiavineyards.com
Email: lnesbitt@chumeiavineyards.com

Clautiere Vineyard
1340 Penman Springs Road
Paso Robles, CA. 93446
P 805-237-3789
F 805-237-1730
URL www.clautiere.com
Email info@clautiere.com

Clayhouse Vineyard
179 Niblick Road PMB 332
Paso Robles, CA. 93446
P 805-239-8989
F 805-238-7247
URL www.clayhousewines.com

Dark Star Cellars
2985 Anderson Road

Paso Robles, CA. 93446
P 805-237-2389
F 805-237-2589
URL www.darkstarcellars.com

Denner Vineyards and Winery
5414 Vineyard Drive
Paso Robles, CA. 93446
P 805-239-4287
F 805-239-0154
URL www.dennervineyards.com

Doce Robles
2023 Twelve Oaks Drive
Paso Robles, CA. 93446
P 805-227-4766
F 805-227-6521
Email: docerobleswinery@tcsn.net

Donati Family Vineyard
2720 Oak View Road
Templeton, CA. 93465
P 805-238-0676
F 805-238-9257
URL www.donatifamilyvineyard.com
Email info@donatifamilyvineyard.com

Dover Canyon Winery
4250 Vineyard Drive
Paso Robles, CA. 93446
P 805-237-0101
F 805-237-9191
URL www.dovercanyon.com
Email dovercanyon@tcsn.net

Dunning Vineyards, Estate Winery and Country Inn
1953 Niderer Road
Paso Robles, CA. 93446
P/F 805-238-4763
URL www.dunningvineyards.com

Eagle Castle Winery
3090 Anderson Road
Paso Robles, CA. 93446
P 805-227-1428
F 805-227-1429
URL www.eaglecastlewinery.com
Email gstemper@tcsn.net

Eberle Winery
PO Box 2459
Paso Robles, CA. 93447
P 805-238-9607
F 805-237-0344
URL www.eberlewinery.com
Email tastingroom@eberlewinery.com

Edward Sellers Vineyard and Wines
PO Box 5141

Paso Robles, CA. 93447
P 805-239-8915
F 805-239-8312
URL www.edwardsellers.com
Email: info@edwardsellers.com

Emerald Hills Vineyard
777 Camino Vina
Paso Robles, CA. 93446
P 303-810-6062
F 720-529-9667
URL www.emeraldhillsvineyard.com
Email ehvine@aol.com

EOS Estate Winery
PO Box 1287
Paso Robles, CA. 93447
P 805-239-2562
F 805-239-2317
URL www.eosvintage.com
Email friends@eosvintage.com

Firestone Vineyard
2300 Airport Road
Paso Robles, CA. 93446
P 805-591-8050
F 805-686-1286
URL www.firestonewine.com

Five Rivers Winery
PO Box 3997
Paso Robles, CA. 93447
P 805-467-0192
F 805-467-2192
URL www.fiveriverswinery.com

Four Vines Winery
3750 Highway 46 West
Templeton, CA. 93465
P 805-237-0055
F 805-227-0863
URL www.fourvines.com
Email info@fourvines.com

Fratelli Perata
1595 Arbor Road
Paso Robles, CA. 93446
P/F 805-238-2809
URL www.fratelliperata.com

Garretson Wine Company
2323 Tuley Court, Suite 110
Paso Robles, CA. 93446
P 805-239-2074
F 805-239-2057
URL www.garretsonwines.com
Email info@garretsonwines.com

Gelfand Vineyards
5530 Dresser Ranch Place

Paso Robles, CA 93446
P 805-239-5808
F 805-239-1507
URL www.gelfandvineyards.com
Email len@gelfandvineyards.com

Graveyard Vineyards
6990 Estrella Road
San Miguel, CA. 93451
P 805-467-2043
F 805-467-2610
URL www.graveyardvineyards.com

Grey Wolf Vineyards and Cellars
2174 Highway 46 West
Paso Robles, CA. 93446
P 805-237-0771
F 805-237-9866
URL www.greywolfcellars.com
Email greywolf@tcsn.net

Halter Ranch Vineyard
8910 Adelaida Road
Paso Robles, CA. 93446
P 805-226-9455
F 805-226-9668
URL www.halterranch.com
Email info@halterranch.com

Hansen Winery
5575 El Pomar
Templeton, CA. 93465
P 805-239-5412
F 805-226-9023
URL www.hansenvineyard.com
Email hansenwines@aol.com

Harmony Cellars
3255 Harmony Valley Road
Harmony, CA. 93435
P 805-927-1625
F 805-927-0256
URL www.harmonycellars.com

Hice Cellars
821 Pine Street, Unit D
Paso Robles, CA. 93446
P 805-237-8888
F 805-237-0181
URL www.hicecellars.com
Email info@hicecellars.com

Hug Cellars
2323 Tuley Ct. Suite 120 D
Paso Robles, CA. 93446
P 805-828-5906
F 805-226-8571
URL www.hugcellars.com
Email augie@hugcellars.com

Hunt Cellars
2875 Oakdale Road
Paso Robles, CA. 93446
P 805-237-1600
F 818-718-8048
URL www.huntwinecellars.com

J. Lohr Vineyards and Wines
6169 Airport Road
Paso Robles, CA. 93446
P 805-239-8900
F 805-239-0365
URL www.jlohr.com

Jack Creek Cellars
5265 Jack Creek Road
Templeton, CA. 93465
P 805-226-8283
F 805-226-8285
URL www.jackcreekcellars.com
Email doug@jackcreekcellars.com

John Alan Winery
605 Moss Lane
Templeton, CA. 93465
P 909-238-7971
F 909-773-0477
URL www.johnalanwinery.com

Justin Vineyards and Winery
11680 Chimney Rock Road
Paso Robles, CA. 93446
P 805-238-6932
F 805-238-7382
URL www.justinwine.com
Email info@justinwine.com

Kenneth Volk Vineyards
281 Broad Street
San Luis Obispo, CA 93401
P 805-782-0425
F 805-781-9489
URL www.volkwines.com

Kukkula
45 Main Street
Templeton, CA. 93465
P 805-227-0111
URL www.kukkulawine.com

Laura's Vineyard
5620 Highway 46 East
Paso Robles, CA. 93446
P 805-238-6300
F 805-238-6911
URL www.laurasvineyard.com
Email info@laurasvineyard.com

Linne Calodo
3845 Oakdale Road

Paso Robles, CA. 93446
P 805-227-0797
F 805-227-4868
URL www.linnecalodo.com
Email info@linnecalodo.com

Lion's Peak
7320 Cross Canyons Road
San Miguel, CA. 93451
P 805-467-2010
F 805-467-3436
URL www.lionspeakwine.com
Email info@lionspeakwine.com

Locatelli Vineyards and Winery
8585 Cross Canyon Road
San Miguel, CA. 93451
P 805-467-0067
F 805-467-0127
URL www.locatelliwinery.com
Email info@locatelliwinery.com

Lone Madrone
At Sycamore Farms on 46 West
P 805-238-0845
URL www.lonemadrone.com

L'Aventure
2815 Live Oak
Paso Robles, CA. 93446
P 805-227-1588
F 805-227-6988
URL www.laventurewinery.com
Email stephanwines@tcsn.net

Madison Cellars
4540 Highway 41
Paso Robles, CA 93446
P 805-237-7544
F 805-237-7798
URL www.madisoncellars.com
Email info@madisoncellars.com

Maloy O'Neill Vineyards
5725 Union Road
Paso Robles, CA. 93446
P 805-238-7320
F 805-226-8412
URL www.maloyoneill.com
Email winery@maloyoneill.com

Martin and Weyrich Winery
PO Box 7003
Paso Robles, CA. 93447
P 805-238-2520
F 805-238-6041
URL www.martinweyrich.com

Meridian
7000 Hwy. 46 East

Paso Robles, CA. 93446
P 805-237-6000
F 805-239-9624
URL www.meridianvineyards.com

Midlife Crisis Winery
1244 Pine Street
Paso Robles, CA. 93446
P 805-237-8730
F 805-237-0109
URL www.midlifecrisiswinery.com
Email jill@midlifecrisis.com

Midnight Cellars
2925 Anderson Road
Paso Robles, CA. 93446
P 805-239-8904
F 805-239-3289
URL www.midnightcellars.com
Email info@midnightcellars.com

Minassian-Young Vineyards
4045 Peachy Canyon Road
Paso Robles, CA. 93446
P 805-238-7571

Mitchella Vineyard and Winery
2525 Mitchell Ranch Way
Paso Robles, CA. 93446
P 805-239-8555
F 805-239-2525
URL www.mitchella.com
Email angela@mitchella.com

Mondo Cellars
3260 Nacimiento Lake Drive
Paso Robles, CA. 93446
P 805-226-2925
URL www.mondocellars.com
Email jmmondo@mondocellars.com

Moonstone Cellars
801 Main St. Suite C
Cambria, CA. 93428
P 805-927-9466
F 805-244-9226
URL www.moonstonecellars.com
Email info@moonstonecellars.com

Nacimiento Cellars
3230 Riverside Ave. Suite 140
Paso Robles, CA. 93446
P 805-226-8100
F 805-226-8188
URL www.nacimientocellars.com
Email ryan@thevintnervault.com

Nadeau Family Vintners
3860 Peachy Canyon Road

Paso Robles, CA. 93446
P 805-239-3574
F 805-239-2314
URL www.nadeaufamilyvintners.com
Email pnadeau@tcsn.net

Nichols Winery and Cellars
P/F 310-305-0397

Niner Wine Estates
1322 Morro Street
San Luis Obispo, CA. 93401
P 805-239-2233
F 805-239-0033
URL www.ninerwine.com
Email info@ninerwine.com

Norman Vineyards
7450 Vineyard Drive
Paso Robles, CA. 93432
P 805-237-0138
F 805-227-6733
URL www.normanvineyards.com

Opolo Vineyards
7110 Vineyard Drive
Paso Robles, CA. 93446
P 805-238-9593
F 805-371-0102
URL www.opolo.com

Orchid Hill Vineyard
2110 Township Road
Paso Robles, CA. 93446
P 805-237-7525
F 805-237-7570
URL www.orchidhillwine.com
Email victor@orchidhillwine.com

Parkfield Vineyards
70502 Vineyard Canyon
San Miguel, CA. 93451
P 805-463-2316
URL www.parkfieldvineyards.com

Peachy Canyon Winery
1480 N. Bethel Road
Templeton, CA. 93465
P 805-237-1577
F 805-237-2248
URL www.peachycanyon.com
Email info@peachycanyon.com

Penman Springs Vineyard
1985 Penman Springs Road
Paso Robles, CA. 93446
P 805-237-8960
F 805-237-8975
URL www.penmansprings.com
Email carlmc@tcsn.net

Pipestone Vineyards
2040 Niderer Road
Paso Robles, CA. 93446
P 805-227-6385
F 805-227-6383
URL www.pipestonevineyards.com

Pomar Junction Vineyard and Winery
110 Gibson Street
Templeton, CA 93465
P 805-434-4100
URL www.pomarjunctionvineyard.com
Email info@mesavineyard.com

Pretty-Smith Vineyards and Winery
13350 N. River Road
San Miguel, CA. 93451
P 805-467-3104
F 805-467-3719
URL www.prettysmith.com
Email info@prettysmith.com

Rabbit Ridge Winery and Vineyards
1172 San Marcos Road
Paso Robles, CA 93446
P 805-467-3331
F 805-467-3339
URL www.rabitridgewinery.com
Hours: Daily 11 a.m. - 5 p.m.

Rainbow's End Vineyard and Winery
8535 Mission Lane
San Miguel, CA. 93451
P 805-467-0044
F 805-467-2304
URL www.rainbowsendvineyard.com
Email rainbow@digitalputty.com

Ranchita Canyon Vineyard
3439 Ranchita Canyon Road
San Miguel, CA. 93451
P 805-467-9448
Email whinrich@yahoo.com

Red Soles Winery
P 805-226-9898
F 805-226-9797
URL www.resoleswinery.com

River Star Vineyards and Cellars
7450 Estrella Road
San Miguel, CA. 93451
P 805-467-0086
F 805-467-2846
URL www.riverstarvineyards.com
Email greatwines@riverstarvineyards.com

RN Estate and Winery
7986 North River Road

Paso Robles, CA. 93446
P 805-467-3106
URL www.rnestate.com
Email rnicolas@rnestate.com

Robert Hall Winery
3443 Mill Road
Paso Robles, CA. 93446
P 805-239-1616
F 805-239-2464
URL www.roberthallwinery.com

Rotta Winery
250 Winery Road
Templeton, CA 93465
P 805-434-9621
F 805-434-9623
URL www.rottawinery.com
Email rottawine@tcsn.net

San Marcos Creek Vineyard
7750 North Highway 101
Paso Robles, CA. 93446
P 866-PASO WINE
F 805-467-0160
URL www.sanmarcoscreekvineyard.com

San Simeon/Maddelena
737 Lamar Street
Los Angeles, CA. 90031
P 323-223-1401
F 323-221-7261
URL www.riboliwines.com
Email info@riboliwines.com

Scott Aaron Wines
422 S. Main Street
Templeton, CA. 93465
P 888-611-9463
URL www.scottaaron.com
Email info@scottaaron.com

Sculpterra Winery and Vineyards
5125 Linne Road
Paso Robles, CA. 93446
P 805-674-8176
F 805-226-8883
URL www.sculpterra.com
Email sculpterrawinery@gmail.com

Silver Horse Winery
2995 Pleasant Road
San Miguel, CA. 93451
P 805-467-9463
F 805-467-9414
URL www.silverhorse.com
Email steve@silverhorse.com

Stacked Stone Cellars
1525 Peachy Canyon Road

Paso Robles, CA. 93446
P 805-238-7872
F 805-238-7495
URL www.stackedstone.com
Email donald@stackedstone.com

Starr Ranch
9320 Chimney Rock Road
Paso Robles, CA. 93446
P 805-227-0144
URL www.starr-ranch.com
Email jwstarr3@earthlink.net

Stephen's Cellar
7575 York Mountain Road
Templeton, CA. 93465
P/F 805-238-2412
URL www.stephenscellar.com
Email steve@stephenscellar.com

Still Waters Vineyards
2750 Old Grove Lane
Paso Robles, CA. 93446
P 805-237-9231
F 805-438-3187
URL www.stillwatersvineyards.com
Email winery@stillwatersvineyards.com

Stolo Family Winery
3770 Santa Rosa Creek Road
Cambria, CA. 93428
P 866-212-7168
F 866-312-4885
URL www.treviti.com
Email stolofamily@gmail.com

Summerwood Winery
2175 Arbor Road
Paso Robles, CA. 93446
P 805-227-1365
F 805-227-1366
URL www.summerwoodwine.com
Email info@summerwoodwine.com

Sylvester Vineyards and Winery
5115 Buena Vista Drive
Paso Robles, CA. 93446
P 805-227-4000
F 805-227-6128
URL www.sylvesterwinery.com
Email info@sylvesterwinery.com

Tablas Creek Vineyard
9339 Adelaida Road
Paso Robles, CA. 93446
P 805-237-1231
F 805-237-1314
URL www.tablascreek.com
Email info@tablascreek.com

Terry Hoage Vineyards
870 Arbor Road
Paso Robles, CA. 93446
P 805-238-2083
F 805-238-2091
URL www.terryhoagevineyards.com
Email jennifer@terryhogevineyards.com

Thacher Winery
8355 Vineyard Ave.
Paso Robles, CA. 93446
URL www.thacherwinery.com
Email info@thatcherwinery.com

Thunderbolt Junction Winery
2740 Hidden Mountain Road
Paso Robles, CA. 93446
P 805-226-9907
F 805-226-9903
URL www.thunderboltjunction.com
Email info@thunderboltjunction.com

Tobin James Cellars
8950 Union Road
Paso Robles, CA. 93446
P 805-239-2204
F 805-239-4471
URL www.tobinjames.com

Tolo Cellars
9750 Adelaida Road
Paso Robles, CA. 93446
P 805-226-2282
F 805-226-2292
URL www.tolocellars.com
Email josh@tolocellars.com

Treana Winery
PO Box 3260
Paso Robles, CA. 93447
P 805-238-6979
F 805-238-4063
URL www.treana.com
Email info@treana.com

Turley Wine Cellars
2900 Vineyard Drive
Templeton, CA. 93465
P 805-434-1030
F 805-434-4279
URL www.turleywinecellars.com
Email tasting@turleywinecellars.com

Venteux Vineyards
1795 Las Tablas Road
Templeton, CA. 93465
P 805-369-0127
F 805-434-9739
URL www.venteuxvineyards.com
Email info@venteuxvineyards.com

Victor Hugo Winery
2850 El Pomar
Templeton, CA. 93465
P 805-434-1128
F 805-434-1124
URL www.victorhugowinery.com
Email sales@victorhugowinery.com

Vihuela Winery
995 El Pomar
Templeton, CA; 93465
P 805-886-5400
URL www.vihuelawinery.com
Email mike@pecen.org

Villa Creek Cellars
1144 Pine Street
Paso Robles, CA. 93446
P 805-238-7145
F 805-238-7145
URL www.villacreek.com
Email wine@villacreek.com

Villicana Winery
2725 Adelaida Road
Paso Robles, CA. 93446
P 805-239-9456
F 805-239-0115
URL www.villicanawinery
Email villicanawinery@earthlink.net

Vina Robles
PO Box 699
Paso Robles, CA. 93446
P 805-227-4812
F 805-227-4816
URL www.vinarobles.com
Email m.ladderiere@vinarobles.com

Vista Creek Cellars
729 13th Street
Paso Robles, CA. 93446
P 805-610-1741
F 805-226-2048
URL www.vistacreekcellars.com
Email info@vistacreekcellars.com

Vista Del Rey Vineyards
7340 Drake Road
Paso Robles, CA. 93446
P 805-467-2138
F 805-467-2765

Whalebone Winery
8325 Vineyard Drive
Paso Robles, CA. 93446
P 805-239-8590
F 805-237-1684
URL www.whalebonevineyard.com

Email info@whalebonevineyard.com

Wild Coyote
3775 Adelaida Road
Paso Robles, CA. 93446
P 805-610-1311
F 805-239-4770
URL www.wildcoyote.biz
Email info@wildcoyote.biz

Wild Horse Winery
1437 Wild Horse Winery Court
Templeton, CA. 93465
P 434-2541
F 805-434-3516
URL www.wildhorsewinery.com
Email info@wildhorsewinery.co

Windward Vineyard
1380 Live Oak Road
Paso Robles, CA. 93446
P 805-239-2565
F 805-239-4005
URL www.windwardvineyard.com
Email marcgoldberg@windwardvineyard.com

York Mountain Winery
7505 York Mountain Road
Templeton, CA. 93465
P 805-238-3925
F 805-238-0428
URL www.yorkmountainwinery.com

Zenaida Cellars
1550 Hwy. 46 West
Paso Robles, CA. 93446
P 805-227-0382
F 805-227-0349
URL www.zenaidacellars.com
Email info@zenaidacellars.com

APPENDIX D

- Map of proposed Paso Robles AVA Districts;

- Map of Paso Robles AVA Major Watersheds;

- Map of North County wineries and tasting rooms west of Highway 101; and

- Map of North County wineries and tasting rooms east of Highway 101.

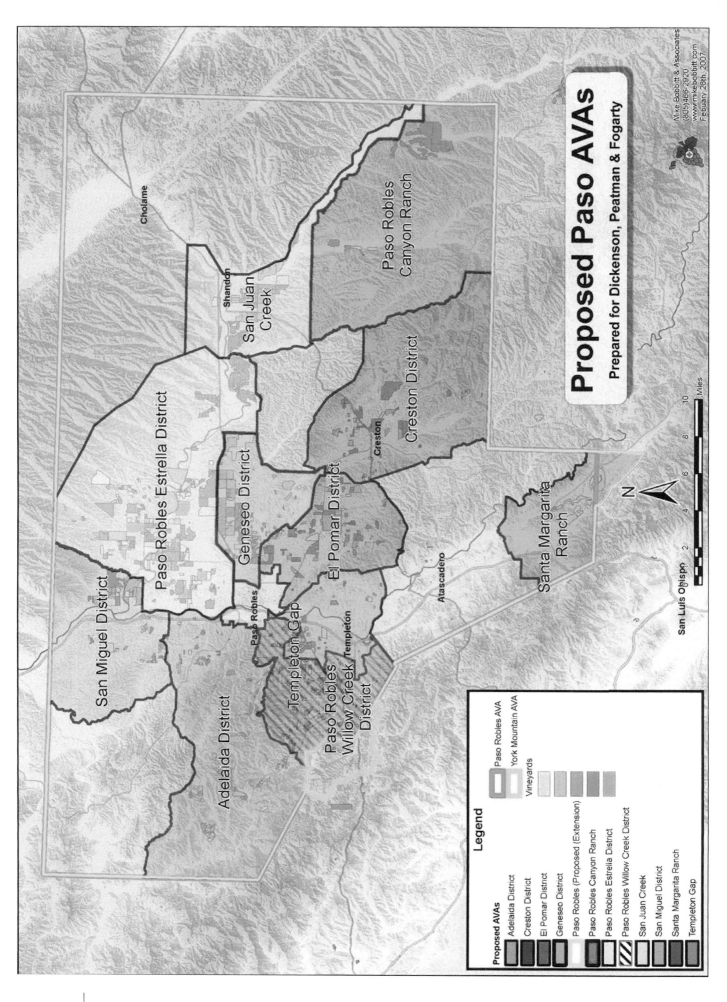

Proposed Paso AVAs

Prepared for Dickenson, Peatman & Fogarty

Mike Bobbitt & Associates
(805)466-2920
www.mikebobbitt.com
Febuary 26th, 2007

N

Miles
2 4 6 8 10

Legend

Proposed AVAs

Adelaida District
Creston District
El Pomar District
Geneseo District
Paso Robles (Proposed (Extension)
Paso Robles Canyon Ranch
Paso Robles Estrella District
Paso Robles Willow Creek District
San Juan Creek
San Miguel District
Santa Margarita Ranch
Templeton Gap

Paso Robles AVA
York Mountain AVA

Vineyards

San Miguel District

Adelaida District

Paso Robles Estrella District

Geneseo District

Paso Robles Willow Creek District

Templeton Gap

El Pomar District

Creston District

San Juan Creek

Paso Robles Canyon Ranch

Santa Margarita Ranch

Cholame

Shandon

Paso Robles

Templeton

Creston

Atascadero

San Luis Obispo

Paso Robles AVA Major Watersheds

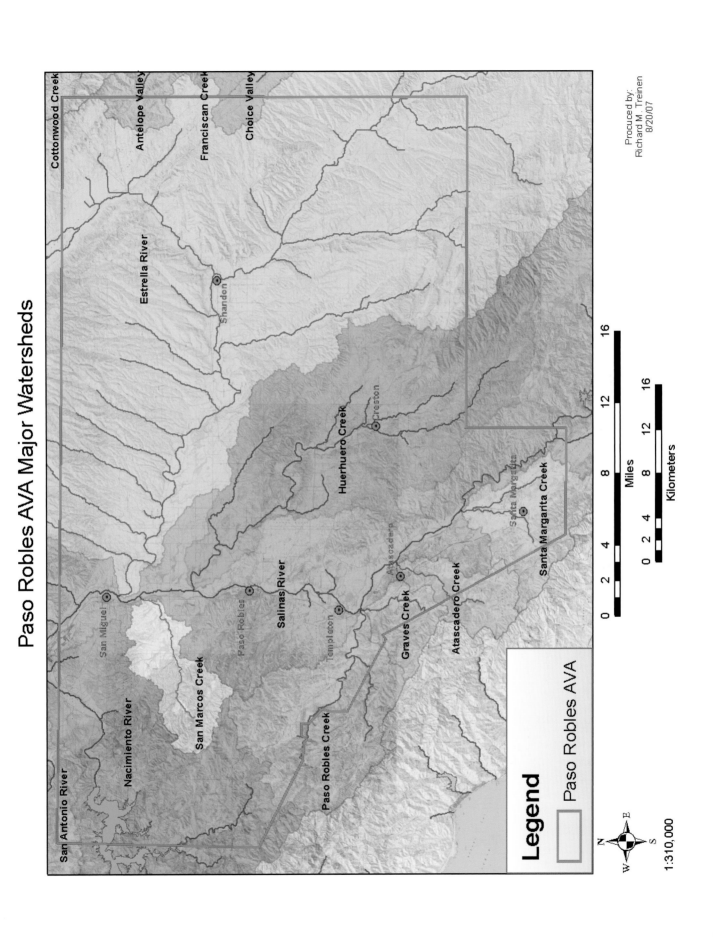

Procued by:
Richard M. Treinen
8/20/07

Cottonwood Creek

Antelope Valley

Franciscan Creek

Choice Valley

Estrella River

Shandon

San Antonio River

Nacimiento River

San Marcos Creek

San Miguel

Paso Robles

Salinas River

Huerhuero Creek

Preston

Paso Robles Creek

Templeton

Atascadero

Graves Creek

Atascadero Creek

Santa Margarita

Santa Margarita Creek

Legend

Paso Robles AVA

1:310,000

0 2 4 8 12 16 Miles

0 2 4 8 12 16 Kilometers

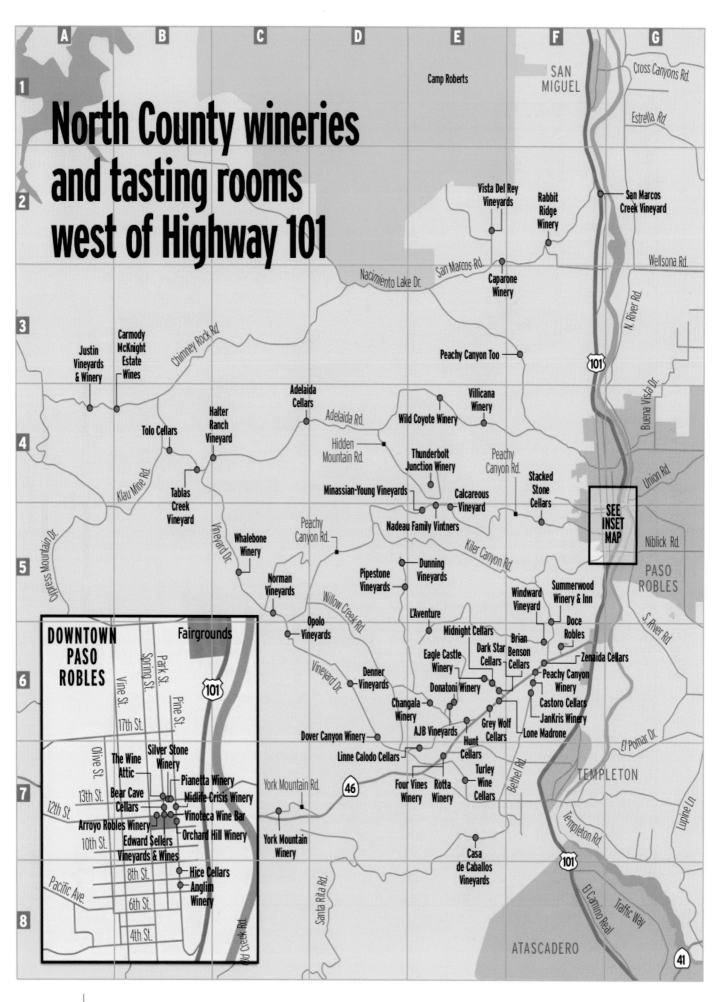

North County wineries and tasting rooms west of Highway 101

A **B** **C** **D** **E** **F** **G**

1

Camp Roberts

SAN MIGUEL

Cross Canyons Rd.

2

Vista Del Rey Vineyards

Rabbit Ridge Winery

San Marcos Creek Vineyard

Estrella Rd.

San Marcos Rd.

Caparone Winery

Nacimiento Lake Dr.

Wellsona Rd.

N. River Rd.

3

Carmody McKnight Estate Wines

Justin Vineyards & Winery

Chimney Rock Rd.

Peachy Canyon Too

Buena Vista Dr.

101

4

Adelaida Cellars

Adelaida Rd.

Villicana Winery

Wild Coyote Winery

Union Rd.

Halter Ranch Vineyard

Tolo Cellars

Hidden Mountain Rd.

Thunderbolt Junction Winery

Peachy Canyon Rd.

Stacked Stone Cellars

SEE INSET MAP

Klau Mine Rd.

Tablas Creek Vineyard

Minassian-Young Vineyards

Calcareous Vineyard

Niblick Rd.

Cypress Mountain Dr.

Vineyard Dr.

Whalebone Winery

Peachy Canyon Rd.

Nadeau Family Vintners

Kiler Canyon Rd.

PASO ROBLES

5

Norman Vineyards

Pipestone Vineyards

Dunning Vineyards

Windward Vineyard

Summerwood Winery & Inn

S. River Rd.

Willow Creek Rd.

L'Aventure

Doce Robles

Opolo Vineyards

Midnight Cellars

Brian Benson Cellars

Zenaida Cellars

6

Denner Vineyards

Eagle Castle Winery

Dark Star Cellars

Peachy Canyon Winery

Vineyard Dr.

Changala Winery

Donatoni Winery

Castoro Cellars

JanKris Winery

El Pomar Dr.

Dover Canyon Winery

AJB Vineyards

Grey Wolf Cellars

Lone Madrone

Linne Calodo Cellars

Hunt Cellars

TEMPLETON

7

York Mountain Rd.

46

Four Vines Winery

Rotta Winery

Turley Wine Cellars

Bethel Rd.

Templeton Rd.

Lupine Ln.

York Mountain Winery

Santa Rita Rd.

Casa de Caballos Vineyards

101

El Camino Real

Traffic Way

8

Old Creek Rd.

ATASCADERO

41

Downtown Paso Robles (inset)

DOWNTOWN PASO ROBLES

Fairgrounds

Park St.

Spring St.

Vine St.

Pine St.

101

17th St.

Olive St.

The Wine Attic

Silver Stone Winery

13th St.

Bear Cave Cellars

Pianetta Winery

12th St.

Midlife Crisis Winery

Vinoteca Wine Bar

Arroyo Robles Winery

Orchard Hill Winery

10th St.

Edward Sellers Vineyards & Wines

Pacific Ave.

8th St.

Hice Cellars

6th St.

Anglim Winery

4th St.

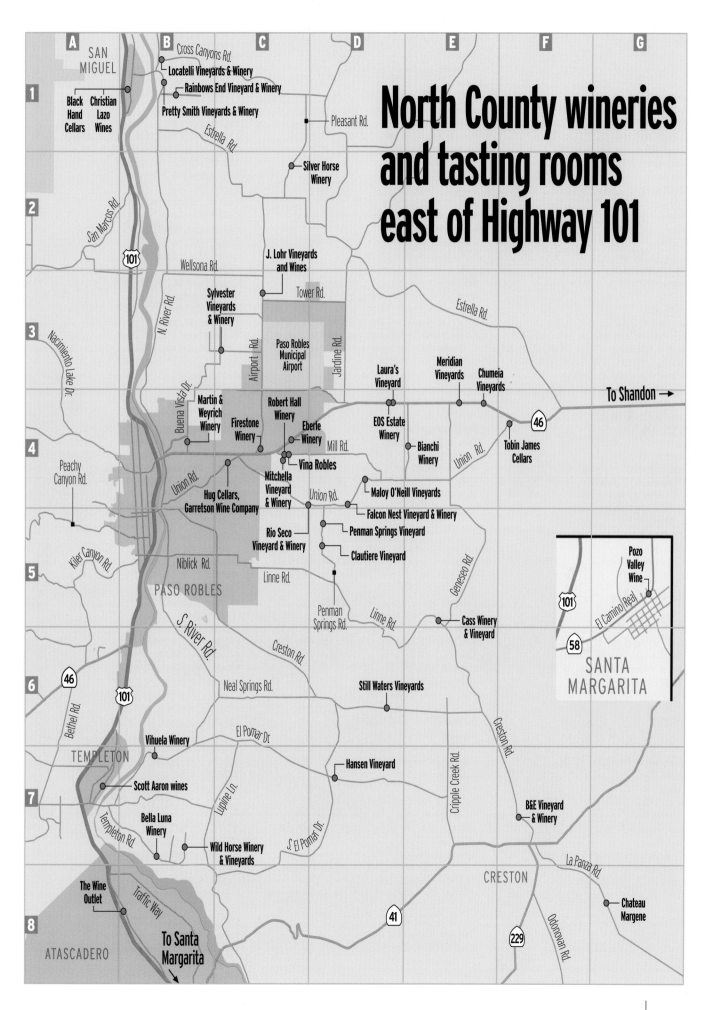

North County wineries and tasting rooms east of Highway 101

SAN MIGUEL

Cross Canyons Rd.

Locatelli Vineyards & Winery

Rainbows End Vineyard & Winery

Pretty Smith Vineyards & Winery

Black Hand Cellars

Christian Lazo Wines

Pleasant Rd.

Estrella Rd.

San Marcos Rd.

101

Silver Horse Winery

Wellsona Rd.

N. River Rd.

J. Lohr Vineyards and Wines

Tower Rd.

Sylvester Vineyards & Winery

Airport Rd.

Paso Robles Municipal Airport

Jardine Rd.

Estrella Rd.

To Shandon →

Nacimiento Lake Dr.

Buena Vista Dr.

Martin & Weyrich Winery

Robert Hall Winery

Firestone Winery

Eberle Winery

Mill Rd.

Laura's Vineyard

EOS Estate Winery

Meridian Vineyards

Chumeia Vineyards

46

Vina Robles

Bianchi Winery

Union Rd.

Tobin James Cellars

Peachy Canyon Rd.

Mitchella Vineyard & Winery

Hug Cellars, Garretson Wine Company

Union Rd.

Union Rd.

Maloy O'Neill Vineyards

Falcon Nest Vineyard & Winery

Rio Seco Vineyard & Winery

Penman Springs Vineyard

Clautiere Vineyard

Kiler Canyon Rd.

Niblick Rd.

PASO ROBLES

Linne Rd.

Penman Springs Rd.

Linne Rd.

Geneseo Rd.

Pozo Valley Wine

101

El Camino Real

S. River Rd.

Creston Rd.

Cass Winery & Vineyard

58

SANTA MARGARITA

46

101

Bethel Rd.

Neal Springs Rd.

Still Waters Vineyards

Creston Rd.

Vihuela Winery

El Pomar Dr.

TEMPLETON

Hansen Vineyard

Cripple Creek Rd.

Scott Aaron wines

Lupine Ln.

B&E Vineyard & Winery

Bella Luna Winery

S. El Pomar Dr.

La Panza Rd.

Wild Horse Winery & Vineyards

CRESTON

The Wine Outlet

Traffic Way

To Santa Margarita

41

229

Odonovan Rd.

Chateau Margene

Templeton Rd.

ATASCADERO

Iron (Fe), 15, 36, 95, 130, 136, 144, 145, 146, 148.

Irrigation, 39, 41, 42, 69, 80, 85, 97, 100, 112, 115, 117, 118, 125, 127, 130, 136, 138, 141, 143, 146, 148, 151, 153.

Italy, 16, 17, 43, 45, 46, 67, 69, 73, 89, 98, 102.

J

J. Lohr Vineyards & Winery, 11, 59, 97, 98, 107, 159.

Jacobsen, Jim and Maribeth, 68, 81, 85.

James Berry vineyard, 70.

James, Joanne, 47.

James, Tobin, 114, 162.

Jardine Vineyard, 115.

Jenson, Josh, 55.

Jimmy's Vineyard, 81.

Juan de Fuca Plate, 29.

Jurassic, 37.

JUSTIN Vineyard and Winery, 17, 59, 60, 61, 63, 88, 116, 151, 159.

K

Kahler, Justin, 76.

Katherine's Vineyard, 70.

Kerman, 95.

Kiler Canyon Vineyard, 88.

Kilmer (soils), 6, 12.

King, Dave and Carol, 51, 52, 66, 154.

Kinne, Alan, 102.

Kleck, Dan, 49, 106.

Klintworth, Gerd, 2.

Kolb Vineyards, 111.

Korecki, Jon and Margie, 119.

Kotze, Lood, 106.

Kroener, Steve, 51.

L

Lagrein, 73.

Lane, Tom, 95.

La Panza Range, 20, 26, 118, 120, 121.

Laura's Vineyard, 100, 101, 102, 153, 159.

L'Aventure Winery, 33, 63, 85, 86, 87, 88, 141, 151, 159.

Liberty School Winery, 71, 73.

Limestone, 2, 34, 35, 37, 38, 55, 76, 80, 127, 129, 130, 131, 137, 143.

Linne (soils), 2, 6, 8, 9, 18, 19, 20, 27, 36, 127, 128, 129, 130, 131, 132, 133, 142, 146.

Linne Calodo Cellars, 51, 88, 89, 159.

Lion's Peak Winery, 45, 159.

Loam, 4, 7, 10, 12, 95, 96, 97, 137, 138, 139, 140, 142.

Loamy sand, 7, 9, 10, 15, 136, 137, 138, 139, 140.

Locatelli Vineyards & Winery, 45, 46, 159.

Locatelli, Cesare, 45.

Locatelli, Louis Gregory, 46.

Locatelli, Raynette, 46.

Lock Ranch Vineyard, 88.

Lockwood (soils), 7, 18, 19, 20, 142.

Lodo (soils), 18.

Lonesome Oak Vineyard, 96.

Longitude, 22.

Lohr, Jerry, 2, 97, 154.

Lompico (soils), 9, 16, 22, 27.

Lopez (soils), 9, 18, 19, 20, 23, 142.

Los Angeles, 1, 48, 51, 69, 89, 109, 161.

Los Osos (soils), 6, 7, 9, 18, 19, 20.

M

MacGillivray, 56.

Macgregor Vineyard, 70.

Madison Cellars, 119, 159.

Magnesium (Mg), 4, 36, 97, 127, 130, 138, 144, 145, 146, 152.

Mahler, Susan, 83.

Malvasia Bianca, 73, 109.

Malbec, 46, 51, 59, 73, 81, 85, 90, 97, 103, 105, 107, 115, 117, 118, 119.

Malolactic fermentation, 73.

Maloy O'Neill Vineyards, 109, 159.

Manganese (Mn), 34, 130, 144, 145.

Marietta, Dominic, 66.

Maritime, 22, 43, 53, 66, 76, 94, 95, 118.

Marsanne, 3, 63, 71, 85, 102, 106, 116.

Martin Brothers, 69, 102.

Martinelli, 2.

Martin, Nick, 102.

Martin and Weyrich Winery, 102, 103, 159.

Mastantuono, Pasquale, 89.

Mastantuono Winery, 89.

Master Water Plan, 40, 41.

Matilija Vineyard, 36.

McCasland, Carl and Beth, 111.

McClean, Mike and Judy, 116.

McClean Vineyards, 116, 117, 154.

McMullin (soils), 16, 20, 22.

Mediterranean, 1, 15, 16, 17, 22, 26, 47, 54, 66, 73, 85, 99, 127, 148.

Melendez, Leslie, 98.

Meridian Winery, 2, 49, 103, 107, 159.

Merlot, 17, 45, 46, 48-51, 55, 57, 59, 62, 64, 65, 67-73, 75, 77, 81, 83, 85, 89, 90, 93, 95, 97-99, 103-105, 108, 109, 111, 112, 114, 115, 117, 118, 119, 123.

Mer Soleil Vineyard, 71.

Mesa Vineyard Management, 67.

Mesozoic, 37, 149.

Messer, Erika, 76.

Messer, Lloyd, 76.

Metamorphic rocks, 29, 32, 35, 36, 38, 147.

Metz (soils), 6, 7, 10, 12, 20, 27.

Meyers, Tom, 67.

Micaceous, 36, 136.

Michel, Hans, 115.

Midlife Crisis Winery, 109, 111, 154, 159.

Midnight Cellars, 90, 160.

Mill Road Vineyard, 96.

Millsholm (soils), 9, 18, 22, 27.

Mineralogy class (soils), 14.

Miocene, 1, 4, 20, 30, 31, 34, 35, 36, 37, 79, 127.

Mission View Estate Winery, 46.

Mittan, Kevin and Jill, 109.

Mocho (soils), 6, 10, 12, 18, 19.

Moderately well drained (soils), 10.

Moisture regime (soils), 14.

Mollic epipedon, 15.

Mollisols, 14, 16, 17, 18, 20, 106, 116, 130.

Montalcino, 43.

Montmorillonite, 16, 17, 55, 130.

Monterey Bay, 22, 23.

Monterey County, 39, 51, 95.

Monterey Formation, 4, 8, 15, 17, 20, 21, 31, 32, 33, 34, 35, 37, 38, 53, 55, 66, 76, 79, 81, 82, 86, 116, 127, 148.

Mooney, Chris, 118.

Mooney, Jon, 118.

Mooney, Michael and Margene, 118.

Moore Warren, 3, 151.

Morgan, Scott, 77.

Morgan, Tom and Sheila, 76, 77.

Mountain(s), 1, 4, 6, 7, 10, 12, 15, 22, 23, 26, 27, 29, 36, 42, 52, 53, 66, 76, 81, 91, 94, 95, 103, 118, 122, 149, 152.

Mourvedre, 17, 44, 53, 60, 63, 64, 70, 71, 73, 80, 81, 84, 85, 88, 89, 92, 99, 100, 105, 106, 107, 115.

Mudstone, 7, 8, 9, 15, 16, 17, 21, 30, 31, 35, 37, 38, 53, 56, 76, 80, 86, 88, 127, 129, 130, 131.

Munch, John, 53, 63.

Muscat Blanc, 53.

Muscat Canelli, 67, 77, 89, 90, 91, 93, 96, 109, 112.

Mustang Springs, 69, 70.

Mustard Creek, 7, 69.

N

Nacimiento Fault, 20.

Nacimiento (soils), 6, 7, 8, 9, 10, 18, 19, 20, 21, 26, 28, 39, 57, 127, 128, 129, 130, 131, 132, 133, 134.

Nacimiento (Lake) River, 26, 28, 39, 70, 148, 159.

Nadeau Family Vintners, 61, 160.

Nadeau, Robert and Patrice, 61, 62, 70.

Nebbiolo, 43, 49, 50, 62, 68, 69, 98, 102, 103, 111.

Nef, Hans, 115.

Nematodes, 27, 28, 117, 141, 145.

Nerelli, Frank, 61, 66.

Nesbitt, Lee, 107.

Nesbitt, Mark, 107, 108.

Newkirk, Cindy and Tim, 112, 113, 114.

New Zealand, 59, 152.

Nicolas, Roger, 105, 161

Nichols, Dave, 90.

Nichols Winery and Cellars, 160.

Niner, Dick and Pam, 103, 104.

Niner Wine Estates, 103, 160.

Nitrates, 39, 138.

Nitrogen (N), 4, 77, 130, 133, 134, 141, 143, 144, 145, 148, 151, 152, 153.

Norman Vineyards, 62, 80, 160.

Norman, Art and Lei, 62.

North American Plate, 16, 29.

O

Oak, American, 46, 49, 50, 51, 52, 54, 55, 61, 62, 68, 69, 70, 71, 76, 77, 81, 88, 91, 93, 95, 98, 99, 105, 116, 119, 123.

Oak, French, 45, 46, 49-52, 54, 55, 57, 61-65, 68, 69-73, 75, 76, 77, 80, 83, 84, 85, 88, 90-93, 95, 97, 98, 99, 104, 105, 108, 109, 116-119, 123.

Oak, Hungarian, 46, 51, 61, 73, 77, 84, 91, 104, 105, 108, 109, 112, 117.

Oak, Russian, 51.

Oakdale Road, 88, 89, 158, 159.

Oak root fungus, 28.

Obispo Formation, 31, 32, 34, 37.

Ochrepts, 15, 18.

Ochric epipedon, 15.

Ogorsolka, Eric, 75.

Old School House Vineyard, 69, 70.

Oligocene, 31, 36, 37.

Olive, 3, 48, 55, 119, 120.

Olmo, Harold, 43.

O'Neill, Shannon, 109.

Opolo Vineyards, 90, 91, 160.

Orange Muscat, 59, 60.

Organic (viticulture), 56, 63, 66, 67, 68, 80, 91, 97, 122.

Organic matter (soils), 5, 14-17, 20, 22, 26, 27, 130, 133, 134, 136, 138, 140, 143, 145.